RIVERS OF OIL

RIVERS OF OIL

**The Founding of
North America's
Petroleum Industry**

HOPE MORRITT

QUARRY PRESS

The publisher gratefully acknowledges the assistance of The
Canada Council, the Ontario Arts Council, the Department of
Communications, and the Ontario Publishing Centre. The pub-
lisher also thanks Donna McGuire and the staff of the Oil
Museum of Canada for their direction.

Canadian Cataloguing in Publication Data

Morritt, Hope
        Rivers of oil: the founding of North America's petroleum
industry

Includes bibliographical references
ISBN 1–55082–088–5

        1. Petroleum industry and trade — Ontario — Oil
Springs — History.   I. Title

HD9574.C23056 1993   C93–090520–2
338.2'7282'0971327

Cover photograph courtesy of the Archives of Ontario.
Design consultant: Keith Abraham.
Printed and bound in Canada
by Webcom Limited, Toronto, Ontario.

Published by
Quarry Press, Inc.,
P.O. Box 1061, Kingston, Ontario K7L 4Y5.

CONTENTS

ACKNOWLEDGEMENTS

I am thankful for the many friends who helped me put this book together. To name a few: Ed. Phelps, librarian at Weldon Library, the University of Western Ontario, loaned me his M.A. Thesis on John Henry Fairbank; Ed. was also very patient in answering questions concerning historical events, either by phone or letter. Author Norma West Linder and historian Jean Turnbull Elford read the manuscript to keep me on track with regard to grammar, syntax, and history. Friedhilde Zilliges and Brother Peter Webbers patiently translated from German to English the many documents on William H. McGarvey who became famous and wealthy in Austria; Dr. Hermann F. Spoerker, former director Osterreichische Mineralolverwaltung, Vienna, Austria sent me copies of documents concerning McGarvey's life in Galicia. Author/artist Peggy Fletcher helped me with art work on early pictures. Historian John Drage supplied me with anecdotes on pioneer life. Brother Patrick Smit was always available when I had problems with a new word processor. Charles O. Fairbank, great grandson of the pioneer John Henry Fairbank, read the final manuscript to make sure that mention of early technology was correct; Charlie today operates the old pioneer field that was his great grandfather's, with four hundred old wells still producing. Dan Cameron, my husband, who worked in oil fields in Canada and Libya for thirty years, also helped me with the technology. My son, Ian Michael Cameron, and daughter, Lynn Cameron Gravel, offered much needed encouragement when my spirits sagged. And my heartfelt thanks go to Bob Hilderley of Quarry Press for his untiring efforts at putting my story of these amazing pioneers into book form.

FOREWORD

One Sunday in the 1960s, my husband and I, with our two small children, drove 27 miles southeast from Sarnia to the village of Oil Springs, Ontario. Friends had told us to make this trip, "To see the first oil well drilled on this continent." We were new residents of Southwestern Ontario, having arrived a few months earlier from Edmonton.

When we reached the pioneer oil field just outside of Oil Springs, I was surprised at the small, ash-pole derricks that were spread out across the land like bleached bones. I had been used to seeing steel derricks that reached into the clouds at Leduc, Alberta where oil was discovered in 1947. Several of the old wells were still pumping, the cross beams nodding lazily in the noon-day sun. Jerker lines — double lengths of wooden rods — hooted softly as they moved back and forth, drawing oil from nearby wells.

I love history, and I wanted to know all about the pioneers who walked these fields. I had worked in the Yukon a few years earlier and had studied the Klondike gold rush of 1898. But this pioneer oil field was the site of another stampede forty-eight years before the Klondike. Why had I not heard about it — or read about it?

An oil museum was located on the site of the first oil well in North America, and I spoke to the curator, asking for books on the early oil pioneers. He informed me that there were no books . . . just articles here and there, but I'd have to hunt them down at a local library. I walked around the museum and looked at pictures on the walls — James Miller Williams, looking prosperous, well-dressed, who first struck oil in North America in 1858; Charles Nelson Tripp, a mercurial mineral explorer who sup-

plied asphalt to pave the streets of Louis Napoleon's Paris; Jake Englehart, a whiskey drummer in New York who founded Imperial Oil, Ltd. in London, Ontario; John Henry Fairbank, a penniless immigrant who struck it rich and built a royal castle in Petrolia to placate a neurotic wife; William H. McGarvey, the first reeve of Oil Springs who struck it rich in the oil fields of Galicia in the Austrian Empire; and other "hard oilers" and oil barons who founded this international industry.

Where did these people originate? How did they hear about the oil bonanza? The curator did not know the answers to these questions — and so began my interest in the early oil history of Canada.

Canadians are slow to praise their heroes, and nothing proves it more than the story of the early oil men who launched an industry in this remote corner of Ontario. Few Canadians — and certainly fewer Americans — know the names of these heroes; no book has put them together in the same way that the life and times of the early oil men of Pennsylvania have been documented. Their lives are scattered in diverse articles, early census records, newspaper accounts, fragments of letters and diaries where their hopes, dreams, fears, and disappointments are hidden, too. *Rivers of Oil* tells the story of these heroes, the first to discover oil in North America. It is a story of greed and avarice, hate and love, violence and shattered dreams, but also a story of vast riches and the struggle of oil pioneers to control a wayward infant.

Canada has a rich oil history, dating back to 1852 when a ruggedly-handsome young man named Charles Nelson Tripp probed through tarry residues in a swamp near Sarnia, Ontario to manufacture asphalt. Five years later, James Miller Williams, a bewhiskered, stubborn carriage maker from Hamilton, cut a swathe in the earth in this same area and found oil. By 1858, Williams was selling his crude and refined product in Canada and the U.S.A. But this story is not well-known by Canadians and hotly disputed by Americans who lay claim to the "first" oil well.

A school teacher in Vancouver recently asked her grade seven class: "Could any student tell me where oil was first discovered in North America?"

A tousle-haired lad waved his hand. "In Pennsylvania," he said. "A guy named Drake . . . in 1859. I know, because my Dad told me."

A teacher in Halifax asked his class: "Where was the first oil well in Canada located?"

A dozen students raised their hands, and one answered for the group: "In Leduc, Alberta . . . 1947."

In 1960, a Canadian geologist, writing to an oil executive in the United States, stirred up a storm of protest when he declared that Oil Springs, Ontario was the site of the first commercial oil well in North America. R. Bruce Harkness, who had worked for the Ontario Department of Mines as a gas commissioner for thirty years, also said that, in 1852, Charles Nelson Tripp, working in Oil Springs, was the first person to use asphalt as a viable, commercial product, and James Miller Williams the first to sink a well, refine the oil from that well, and sell kerosene on a commercial basis, thus launching an industry. A scholarly man, well-versed in the history of oil, Harkness also debunked stories circulated by American historians, that Col. Edwin L. Drake was the "Father of the Oil Industry" in North America, and Titusville, Pennsylvania was the site of "the first commercial oil field" on this continent.

The vice-president of a Pennsylvania oil company, corresponding with Harkness, loudly insisted that in 1859 Titusville *did* become the first oil field, and Col. Drake *was* the "Father of the Oil Industry" on this continent. He angrily asked why, one hundred years after oil had been discovered in North America, were Canadians suddenly claiming to have launched the petroleum industry?

Harkness answered, impatiently: "Well — Canadians are not given to Jingoism. I knew of this [first oil well] twenty-five years ago, and your own prominent American geologist, Frederick G. Clapp, who was engaged by the Canadian government to prepare a report on the Natural Gas and Petroleum Resources in 1914–15, says, 'In fact, Williams in 1857 drilled a deep well in Ontario with successful results before Drake's lucky find.' "[1]

Even if they do know their oil history, Canadians are certainly *not* "given to Jingoism," and nothing proves it more than the story of Tripp, Williams, and other early oil men who launched the petroleum industry.

These men were, indeed, princes of their realm. They found a rich harvest in a lonely swamp; they became lords of all they surveyed; a few arrived as penniless immigrants and became multi-millionaires; one joined the ranks of European royalty. And there were others who met death and disaster while struggling to reach the mother lode.

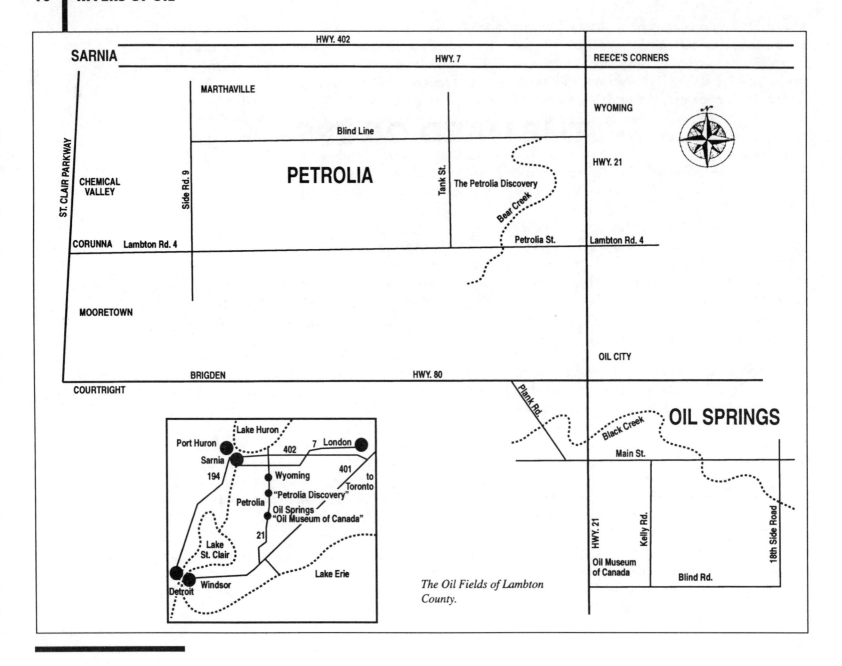

The Oil Fields of Lambton County.

THE HARD OILERS
OF ENNISKILLEN

A BRIEF HISTORY OF OIL

Like gold, oil has lured and mystified man for centuries. In Baku, an ancient city now located near the Iranian border, pillars of flame once licked at the sky where oil bubbled out of the earth. Although these flames were caused by gas leaks, early Persians thought that the gods of fire had produced them, and Zoroastrian priests built a temple at Baku for fire worship. Marco Polo, Italian author and traveler, who lived in the 13th and 14th centuries, made reference in his writings to the pillars of flame at Baku, and also noted that people used oil from the nearby springs to burn in lamps. Indeed, the chief use of oil until the advent of the automobile was for illumination, though some have found more inventive applications.

In the 1st century, a Roman general used oil to confuse his enemy. He located an oil spring, saturated pigs with the greasy substance, torched the animals, and drove them, blazing, into the ranks of his foe. In the First World War, Russians set fire to Austria's vast oil fields in Galicia and created a similar holocaust to strike a crippling blow at Germany's war ships, submarines, and air power. And for centuries, the Indians in Southwestern Ontario used oil from certain oil-soaked lands to cure a whole spectrum of illnesses from headaches to rheumatism. Crude oil was credited with driving evil spirits from tormented persons when lavish amounts were applied to all body parts. If this didn't work, the final cure did: dirty black crude forced down the throat of the ailing one.[1]

As early as 1818, a new form of lighting by gas derived from coal was on the market, but the expense was astronomical. King George IV of England, when he was Prince of Wales, became excited about this new illumination by coal gas, and he spent a fortune to have it installed inside and outside of his extravagant royal residence at Brighton. A company at Black Rock manufactured this gas. Coal was supplied by ships, unloaded on the beach, and hauled up the steep cliffs in buckets by sweating men. George IV was an extravagant man, flamboyant with money, and no cost was too much to light up his Brighton palace. Court guests were terrified of the chandeliers that burned with a constant, loud hissing, and the heat from these fixtures caused many a royal person to faint dead away.

Royalty — and those akin to royalty — were the only people who could afford gas lights, although by the 1850s, a few wealthy Americans and Canadians had installed this kind of light, at great expense, in their homes. Affluent boarding houses in Toronto and New York advertised rooms for rent, with gas lights. Detroit city, in 1851, installed twenty gas lights on downtown streets. Silas Farmer, in his book *The History of Detroit and Michigan*, published in 1884, commented that these lights cost the city $72,000 in one year, and each lamp burned an average of seven hours. Needless to say, the world wanted a cheap illuminating oil.

Whale oil, the most popular lamp oil of the day in 1850, sold for $2.50 a gallon — too expensive for the average household. Also, male whales, from whom the oil was obtained — were close to extinction, and the future of this kind of illuminant looked bleak. Camphene, made from turpentine, was a cheap lamp oil but it was volatile and had sent many a lamp lighter to an early grave. Settlers in Canada and the United States relied on tallow candles or rags dipped in pork or beef grease which gave a foul-smelling, smoky light. People often went to bed when the sun went down and rose when dawn appeared because it was so difficult to work or read by the glow of candles or rags.

In 1854, Dr. Abraham Gesner, a Canadian, patented kerosene. Gesner, at one time a provincial geologist in New Brunswick, extracted oil from natural asphalt beds in Nova Scotia, and by a refining process produced an illuminating oil he called kerosene. He chose the name by combining Greek words *keros* and *elaion*, meaning "wax" and "oil", then changing the spelling of *elaion* to *ene* so that the word would be easier to pronounce.

Kerosene and coal oil are similar substances, but originally the former was retrieved from crude oil, and the latter from coal. Both coal and oil were once part of living organisms, and because they are composed mainly of hydrogen and carbon, they are called hydrocarbons. But oil is a very complex mixture of hydrocarbons. If we were to put crude oil as it comes out of the ground into a tea kettle, put a flame under the kettle and collect, in separate tubes, the successive vapors that pass out the spout, we would accumulate a rough separation of the oil into gases. Indeed, the first refineries, called "stills", were like big tea kettles. If we were able to chill, compress, and absorb the gases that come out of our kettle, we would get first a very light, volatile gasoline, then a denser gasoline, then kerosene; after this the heavier or denser lubricating oils, wax oils, and asphalts would appear.

Dr. Gesner's kerosene was a welcome introduction to the world of the early 1850s, but the oil from which it came was scarce, making it expensive. Canadians continued to light candles or rags dipped in animal fats until Charles Nelson Tripp and James Miller Williams discovered oil in the swamps of Enniskillen Township in the newly formed Province of Canada.

GUM BEDS

In 1850 British North America was a group of scattered territories north of the United States. Ontario, then called Upper Canada or "Canada West", and Quebec, Lower Canada or "Canada East", had been united nine years earlier and called the Province of Canada. All British lands west of the Province of Canada as far as the Pacific Ocean and north to the Arctic Ocean were owned and governed by the Hudson's Bay Company. Nova Scotia, New Brunswick, and Prince Edward Island were sparsely settled provinces in the Maritimes. Confederation, uniting these diverse territories under one banner, was still seventeen years away.

The government of the Province of Canada was the same as that of old Canada West and Canada East — an elected assembly, an appointed council, a governor general, and an executive council. In this system, Canadians were tied to England through the governor general, and radicals like William Lyon Mackenzie in Toronto and Louis-Joseph Papineau in Montreal wanted the American elective pattern that had no yoke with England. The clash between those who wanted British ties and those who wanted the American way of government was one issue that touched off the 1837 Rebellion in the Province of Canada. Although quelled quickly by British troops, the Rebellion caused several hundred rebels to be killed and eight to be exiled to Bermuda; Papineau, Mackenzie, and a few other leaders fled the country.

By 1850, revolutionary elements had quieted down, though. It was time to develop farm lands, build railroads, uncover Canada's vast wealth of minerals.

Nature often guards her treasures behind formidable barriers. Diamonds are imbedded in the ground miles below the earth's surface; gold is locked-in behind remote mountains; pearls are hidden in oysters deep within the sea. The oil in North America's first oil field was no different. Early settlers avoided Enniskillen Township, 82,174 acres of land-locked swamp situated half way between Lake Erie and Lake Huron. The dense forest of walnut, elm, oak, and ash trees was dark and foreboding. Pools of strange, black substances pock-marked the ground, giving off a foul, pungent odor. A sub-soil of blue clay created a quagmire in wet weather, and in the summer months, malaria-carrying mosquitoes rose in clouds from the land. The Sydenham River and its tributaries — Black Creek and Bear Creek — criss-crossed the area, but drainage was poor.

Sir John Colborne, who served as Governor-General of Upper and Lower Canada from 1838 to 1839, named Enniskillen "to honor a Peninsular War colleague of his who had lived in Enniskillen, Ireland."[1] The Canadian territory was boggy and green, much like its namesake, but it was also wild virgin forest, defying settlement, unlike its Irish counterpart.

Enniskillen sat in the center of a larger piece of

land called Lambton County, named for John George Lambton, a British aristocrat who, as Lord Durham, was Governor-General of British North America just prior to Sir John Colborne. The biggest settlement in Lambton in 1850 was Port Sarnia, a fishing and shipping center of 1,000 people nestled on the shore of Lake Huron where the lake flows into the St. Clair River. A ferry boat took passengers and freight across the river, uniting Port Sarnia with Port Huron, a progressive town in Michigan, also devoted to shipping and fishing. Fifty miles down river, the city of Detroit was a booming metropolis of 21,000 residents where foundries, locomotive works, brass factories, saw mills, and ship building flourished.

Port Sarnia was 27 miles northwest of Enniskillen, but it might as well have been a thousand miles away, for roads were non-existent except for a corduroy trail to London that passed within 18 miles of the oil district. Corduroy roads were made of logs, laid at right angles to the road's direction. The Marchioness of Dufferin, the wife of an early governor-general, in describing such a road, said: "When an occasional 'cord' has broken away, when another has got loose and turns around as the horse puts his foot on it, or when it stands up on end as the wheel touches it, the corduroy road is not pleasant to drive many miles over."[2]

London, at one end of this 60-mile log trail, was a growing town of 4,500 people, site of a British military garrison placed there in 1838 to offer protection from rebel forces within the country and any that might cross the international border. Toronto — 120 miles east of London — was a city of 30,000 people, a railroad center jostling with a surge of immigrants who had fled from the potato famine in Ireland. Hamilton, where James Miller Williams lived, was a city situated at the west end of Lake Ontario on Burlington Bay, about 40 miles southwest of Toronto. As a steel center and trans-shipment point located on a newly-built canal, Hamilton was a haven for early oil men.

Although there was great growth in the country with immigrants surging in and railroads flinging their twin-ribbons of steel into the hinterland, there was unrest, too. If one were to draw a line from Toronto to London, Sarnia, Detroit, Hamilton and back to Toronto again, a lopsided rectangle of land would emerge, with the south end straddling the border with the United States. In 1812, at the height of the Napoleonic conflict in Europe, the United States had declared war on Great Britain, and Canada as a British possession felt the impact of battles along the international border. The war made these border communities uneasy, but it also rallied forces within the country, thus forming the beginnings of national identity. United Empire Loyalists — people who wished to remain loyal to the British Crown — had come north after the American Revolution, lured by the offer of free lands and low taxes. They rallied with British troops and civilians to quell American warring factions, and in two years, the war was over.

By 1850, all areas of Lambton County were being settled, except Enniskillen Township. Only thirteen hardy souls were registered as land owners in Enniskillen, and all of these lands were crown grants.[3] As Jean Turnbull Elford says in her book *Canada West's Last Frontier*, "Many of the lots were assigned by the Crown to British soldiers who had served in Canada in the War of 1812, and to those who came out in case they were needed to curb the Rebellion of 1837."[4] Yet, most of the people assigned these Crown lands were absentee landlords as the area was a tangled growth of inaccessible forest and spongy marsh.

Grants in the 1800s were also awarded to the original United Empire Loyalists and their children, as Eric Jonasson, in *The Canadian Geological Handbook*, notes: "Land grants were given to United Empire Loyalists both as a reward for their loyalty to the English Crown, and as a means of populating the country. The extent of the grant varied with the rank which the Loyalist held in the various Loyalist regiments which fought in the American Revolution, the higher ranking officers receiving larger grants than privates. . . . However, the average grant was 200 acres for each head of family, with an additional 200 acres granted to his wife if she was the daughter of a Loyalist, and a further 50 acres for each child in the family who was under age. . . . Over time this was also extended to Loyalists' children, who each received 200 acres on reaching maturity."[5] These free lands were later offered to ordinary settlers, but there was much favoritism shown by the government in the granting of property. Very large tracts of land often went to influential individuals, and a hard-working, sincere, lesser-known person was denied a grant.

Eliakim Malcolm, the first surveyor in Enniskillen in 1832, was not impressed with the area. He said the brushwood and fallen timber were very thick, it was difficult to find adequate drinking water, and, after a snowfall in November, he noted that: "water was from 12 to 18 inches deep. The swamp to the end of the concession is altogether impassable. I therefore have to abandon it [the survey]."[6]

A year later, Lewis Burwell was assigned to make another survey. His boss gave the following instructions: "You will be allowed a party of two chainbearers at three guineas and one shilling six-pence in lieu of ration each per day, and eight axe-men at two shillings six-pence and one shilling six-pence in lieu of ration, and you will be allowed 15 shillings and one shilling six-pence in lieu of ration per day."[7]

On 9 January 1833, Burwell sent in his report to the surveyor-general: "On commencing preparation for the work, I found that no salt pork was to be had in this part of the country nor at any of the villages about the head of Lake Ontario nearer than St. Catharines — and not thinking it safe to risk the chance of getting provisions in the newly settled part of the country where the work was to be performed, I thought it best to go to St. Catharines and purchase my pork, which I found afterwards was the safest method. The quantity of the pork necessary for the work being more than one single team usually carries — and too small for a load for two teams, I however, prevailed upon William Kirby to transport the pork for me with his team by giving him 20 shillings per day. . . . I therefore trust that that item on the account will not be considered an unusual charge." And in his survey report, Burwell concluded that "the streams all take their rise from the swamp, consequently are not durable. In the dry season they are quite dry — but in the wet season discharge great quantities of water. From these circumstances I cannot recommend any of them as suitable for the erection of mills."[8] It was sad that there was no dialogue between these early surveyors and local Indians, for the native people had known about oil and bitumen deposits that they called "gum beds" for many years. As I.C. McCluskey notes, "in 1830, Indians discovered petroleum flowing on the waters of Bear Creek and Black Creek in Lambton County. It was also found in depressions in the ground from which the Indians skimmed it off the water. . . . Refining was not attempted other than by 'boiling down' in open kettles to settle the water more thoroughly, and then filtering through woolen blankets."[9]

Neither Malcolm nor Burwell mention oil seepages or gum beds in their reports, and it took another seventeen years before the "white" man discovered oil. Sterry Hunt, an assistant geologist with the Geological Survey Department of Canada, while tramping through the township, found two beds of bitumen; he reported this bitumen suitable "for the construction of pavements, for paving the bottoms of ships and for the manufacture of illuminating gas."[10]

William E. Logan, the first director of the Geological Survey Department and knighted by Queen Victoria in 1856 for his achievements, was impressed with Hunt's report. A year after Hunt found the bitumen, Logan sent another of his assistants, Alexander Murray, to study further these oil seepages. In 1850, Murray reported that "a visit was made in the early part of the season, to a bed of nearly pure bitumen. . . which, in some parts has the consistency of mineral caoutchouc [rubber]. . . and occurs in the sixteenth lot of the second concession of Enniskillen. . . . The bitumen is underlaid by a very white clay, which I was informed had been bored through in one part for thirty feet."[11]

Logan was delighted; he knew that he and his co-workers had located a ready-made oil field in the swamps of Enniskillen which private entrepreneurs like Charles Nelson Tripp and James Miller Williams would explore and develop.

CHARLES NELSON TRIPP

In 1850, Charles Nelson Tripp, age twenty-seven, was working as foreman of a stove foundry in Bath.[1] Located about 125 miles east of Toronto in Canada West near Kingston, Bath was then a sleepy little village with a few scattered log houses, shops, inns, taverns, and a steamboat wharf.

Born in Schenectady, New York, Tripp was a descendant of Ezekiel Tripp, a staunch Quaker who

Henry Tripp directed his brother Charles Nelson Tripp to the gumbeds of Enniskillen.

moved to the Schenectady district in 1791. His mother was Hanna Clinch whose forbears were inn keepers and staunch members of the Dutch Reform church.

Tripp was considered ruggedly handsome. A Toronto reporter for the *Huron Signal* described him in 1866 as "a remarkable looking man who might be readily taken for a shrewd, sunken-eyed, hard-faced and eccentric western farmer." This reporter rightly entitled his article on Tripp "The Original Oil Man of Canada."[2] He had worked at various jobs in New York before crossing the border into British North America in the late 1840s. He loved adventure — and this love took him into many far-flung areas at a time when train travel, in most places, was still a remote dream. He was creative and industrious. In 1850, he designed a new stove front, patented it, and sold it to John Counter, the owner of the factory where he worked. At that time, he was also interested in mining. He leased and worked lead and copper veins in Hastings and Prince Edward counties near Bath. He envisioned a promising future in minerals for the world and staunchly pursued a dream in which he saw himself both wealthy and famous from the development of mines.

Lead was one of the earliest smelted metals, and lead oxide was used for pottery glazing as early as 7000 B.C. In Tripp's day, there was great need for lead as solder. Copper, alloyed with tin, was once used for weapons and tools by early man. King Solomon's fabled copper mines in the desert near

Elat, Israel, were worked in the 9th century. In the 1850s, copper was alloyed with zinc to make brass utensils, alloyed with tin and zinc for bronze, and alloyed with zinc and nickel for coins and German silver.

It took extensive work and capital to develop both lead and copper claims. The ore had to be mined, crushed, ground, and smelted — no easy feat for a novice. Tripp was probably self-educated with regard to minerals, but all his life he carefully studied every aspect of the mining process. At a later date, friends in New Orleans, Louisiana, said that he knew "more, practically, about the mineral wealth of every southern state, than any other man."[3] All his life, Charles Nelson Tripp searched for big money to satisfy his obsession with mining exploration, but fortune eluded him.

In 1850, his brother Henry, age twenty-two, worked as a plate photographer in Woodstock, about 100 miles southwest of Toronto.[4] He was a shy, quiet young man who seems to have been overshadowed by his exuberant, adventuresome brother. Still, they were close, these two brothers, keeping in touch by mail, and often meeting, person to person, either in Bath or Woodstock. The surveyor Alexander Murray, whose account of the bitumen deposits in Enniskillen appeared in Logan's Report of Progress, 1851–52, also lived in Woodstock.

Woodstock was a small town in those days, a scattering of 2000 people with a tavern, general store, church, one-room school, and hotel. It was the hub of a farming community made up of native-born

Deed for one of many tracts of land purchased by Charles Nelson Tripp in the early 1850s.

Canadians, and recent immigrants from the U. S. A. and British Isles. People knew one another in these small towns, and it is certain that Henry Tripp knew Murray, and Henry was probably one of the first persons to learn of the find in Enniskillen. He moved quickly to inform his brother, for there is evidence in an early *Lambton County Atlas* that Charles Nelson Tripp was located in Enniskillen in 1852, and that year he "erected buildings and machinery for the manufacture of asphalt."[5]

Henry probably helped his brother to get settled, but there is no evidence that he stayed in Enniskillen. In 1855, when Tripp sold 200 acres to Henry, the latter is listed on the land deed as living in Petersburg, Virginia, and Charles Nelson is a resident of Hamilton, Ontario.[6] For the next four years, Tripp visited Hamilton sporadically, buying machinery and drumming up business for his asphalt.

As early as 1853, a year after he took up squatters' rights on Concession 2, Lot 16 (the parcel of land mentioned in Murray's report), he purchased all of these 200 acres for 112 pounds sterling.[7] He was so sure that asphalt was the miracle mineral of the present and future, that, by 1855, he had purchased 1350 acres of prime oil land in Enniskillen.

Tripp was well aware of the expense involved in prospecting, since he had done some exploratory work on lead and copper claims. A few years later, a pioneer wildcatter said that a prospector had to invest "at least $1000 in buying an oil lot, putting down a well and paying overhead before the well was fully operative. . . . Dealers extended ample credit for land and machinery and re-possessed both if the prospector failed to make good."[8] This was the cost and risk of one well on a small parcel of land — perhaps an acre or half an acre. Tripp planned to set up operations for asphalt on 1350 acres.

Rock oil was not new to the world in the 1850s, but the process of quickly and efficiently extracting it from the ground, refining it and marketing the refined product *was* new, and within the next decade, these processes would create a whole new, complicated industry. Venture capital, some knowledge of crude oil, a good grasp of operating costs, roads or railroads to get the product to market, and a lavish amount of luck — these were only a few of the requirements for success. Charles Nelson Tripp had no mentor. He was alone with his 1350 acres and his dreams.

In 1853, he asked Dr. Thomas Antisell, consulting chemist in New York, and Thomas McIlwraith, manager of the Hamilton Gas Company, to analyze samples of asphalt that he had processed from the gum beds on his lands. Antisell reported that the asphalt was "highly suitable for paints, plastics, mastics, waterproofing cements, and as an illuminant by gasification or by distilling an oil from it"; and McIlwraith reported that on using it in place of coal in the gas company's retorts, an excellent illuminating gas was formed.[9]

Tripp did not seem to realize the value of the oil that gurgled underneath his asphalt beds. Instead of probing for a cheap, good illuminant, he stubbornly continued to work with asphalt, seeing it as the answer to sealing the hulls of ships and hard surfacing of roads. In this latter respect, his vision was 20/20. Asphalt was used to pave streets in the U.S.A. as early as 1870. By 1903 thirty-five million square yards of roads were paved with asphalt, and today ninety percent of roads are hard-surfaced with this product. But Tripp, like many geniuses, was years ahead of his time in thinking. He'd been born too soon.

And yet in the early days of his Enniskillen ventures, he had astounding successes. He was scarcely settled on his land when he decided to form a company — The International Mining and Manufacturing Company — which became the first oil company in North America, beating out by four weeks the founding of the Pennsylvania Rock Oil Company in the United States. His petition to the government for a charter said, in part:

Your petitioners have been for the last three years, at great cost and expense, exploring various sections of this province for asphalt, lead, copper, silver, oil and salt springs, and the said Charles N. Tripp is owner in fee of two large asphalt beds in the western district, six oil and two salt springs in the said district, also one lead vein in the County of Prince Edward, lead in the township of Bedford, lead in Belmont, lead and copper in Otonabee, and has mining rights, leases and privileges in various portions of this Province.

That your petitioners have associated themselves together for working and development of the said mines and minerals and for bringing the same into the market, whereby much benefit must result to the Canadian public.[10]

It was also deemed necessary to erect works "for the purpose of making oil naptha, paints, burning fluids, varnishes and other things of the like."[11]

The charter for The International Mining and Manufacturing Company was granted 18 December 1854, although the original petition was sought during the parliamentary session of 1852. Capital for the company was set at 60,000 pounds, and each director was required to subscribe 250 shares, with a par value of five pounds each.

John S. Ewing, in his study of "The History of Imperial Oil," written for Harvard Business School in Boston, comments on the aims of the company and Tripp's business acumen: "These were ambitious ideas. . . . It is difficult to believe that he [Tripp] lacked some training in chemistry and geology, for the mere recital of powers of the company implies some little knowledge of these subjects. He seems also to have had some conception of extensive corporate organization: visionary though his ideas were, they were well in advance of anything in the industrial area then existing in Canada."[12] The directors of The International Mining and Manufacturing Company were Tripp himself; Hiram Cook, age forty-eight, a wood merchant from Hamilton;[13] John B. Van Voorhis, a wood merchant and contractor from Woodstock;[14] and Henry Tripp, who is believed to have been living, at that time, in Petersburg, Virginia.

Application for a Charter for the first oil company in North America, granted 18 December 1854.

In 1854, William E. Logan was assembling a display of Canadian minerals to be shown the following year at the International Exhibition in Paris, France. He asked Tripp to send him asphalt exhibits that he planned to consider for Paris. The government would pay for transportation costs for the asphalt to Montreal, and from there to France. No doubt Tripp was elated at this recognition of his work by a renowned scientist like Logan.

By the end of 1854, life looked good to Charles Nelson Tripp. He had formed an oil company with enthusiastic shareholders, he had raised enough capital to help defray the high costs of wresting asphalt from the ground and marketing it, and his miracle mineral was on its way to France where very important people of the world would see it and would perhaps order many boat loads.

PARIS INTERNATIONAL EXHIBITION

Louis Napoleon — Napoleon 3rd — was on the throne in France when the International Exhibition was held in Paris in 1855. Nephew of Napoleon Bonaparte, the first Emperor of France, Louis Napoleon, at age forty-seven, was a dashing figure of a man, the son of King Louis and Queen Hortense of Holland. A great advocate of agricultural and industrial development, he had masterminded the World Fair, and he made sure that the pomp and splendor of France were on display at the big event.

In 1853, Louis Napoleon was married to Eugenie de Montijo, a Spaniard of great beauty, and together they made the French court famous for its grandeur and its extravagance, while at the same time introducing a new social atmosphere to this country that had known revolution and wars for twenty-six years.

Journalists from leading Canadian newspapers were at the fair, and the dazzle of the event was captured in the *Hamilton Weekly Spectator* in August:

"The Hotel de Ville, in the day decorated with flags, was at night illuminated in all the colors of the rainbow. The Palace and Garden of the Tuilleries, the Arc de Triomphe, the various bridges, the ministers' residences, the Palais Royal and the Universal Exhibition were splendidly illuminated with [gas] lights. . . . On the Boulevard and in the principal streets, a number of houses were tastefully lit up with colored [gas] lamps."[1] The Canadian delegation was also noted by the French. A journalist writing for one of France's leading newspapers, *The Paris Monitor*, noted that "Canada figures admirably at the Exhibition, and its products and its specimens of grains, fruits, flour of all kinds, attract great attention. The care which the commissioners and delegates from Canada have displayed, has merited the just eulogiums which have been addressed to them several times by Prince Napoleon."[2] Another correspondent reported that "Mr Logan's collection of minerals by far eclipses those from any other portions of the world."[3]

George Perry of Montreal, world-renowned for the fire engines he had built, was praised for the display model that won first prize at the 1851 World Fair in London, England. And Logan was lauded again for his minerals: "This gentleman, who is official geologist as well as commissioner to Paris, has exerted himself unremittingly to set Canada off to the best advantage, and assisted by George Perry, has certainly been successful."[4]

William Edmond Logan was an international hero. Born in Montreal in 1798, he was educated at the University of Edinburgh. In 1840, he mapped coal seams in Wales, and his maps were adopted for official use by the Geological Survey of Great Britain. His Canadian birth and renowned position as a geologist won him in 1842 the position of Geologist for the Province of Canada.

Following his impressive display of minerals and his geological map of Canada at the 1851 International Exhibition in England, Logan was honored as the first native-born Canadian inducted into the Royal Society of London. As a result of his high profile, he was constantly on stage at the World Exhibition in Paris. Shortly after the opening ceremonies in May, the Emperor and Empress held a reception at the Palace of the Tuilleries. Here, their Majesties received heads of state and renowned persons from around the world — and Logan was among them.

The sparkle and dazzle of the Exhibition was marred at the beginning when an attempt was made on the life of Louis Napoleon by an Italian named Pianori. The Crimean War also vied for headlines with the International Exhibition. A coalition of Great Britain, France, Sardinia, and Turkey began a major offensive against Russia in 1853. For three years, fierce fighting ensued over control of the straits between the Black Sea and the Mediterranean. In spite of these sad events, the Exhibition continued for six exciting months. At the closing ceremonies in November, William Logan received the Grand Medal of Honor for his exhibits of fifty-one Canadian minerals. Also, the Emperor conferred upon him the title of "Chevalier of the Legion of Honor."[5] Charles Nelson Tripp was awarded an honorable mention for his exhibits of asphalt.[6]

Back in Canada, Tripp was no doubt elated when informed that he received this honor. To gladden his heart further, the City of Paris sent him an order for enough asphalt to pave the main streets of the city. An ultra-modern metropolis, Paris had been in love with asphalt for years. In 1838, it was the first city in the world to build asphalt sidewalks, the bitumen supplied by the French mine Seyssell. At that time the Swiss mine Val-de-Travers was fighting with Seyssell for the right to supply Paris with asphalt for its sidewalks.[7] However, asphalt was scarce, and it seemed that by 1855 the demand was greater than the supply. The city fathers were delighted with the Canadian exhibits at the World Fair, for this could lead them to a plentiful source of asphalt for their roads.

The winter months of 1855-56 must have been busy in Enniskillen, for Tripp had to process his asphalt, then transport it, sleigh load by sleigh load, over frozen land to Port Sarnia, where it was loaded on boats — seven in all — bound for Paris.[8]

While the International Exhibition was in full swing, Tripp met and married Almira Jane Cornish of London, Ontario. There are no records to show how or when they met, but he was traveling in excellent company, for eighteen-year-old Almira was the daughter of Dr. William King Cornish, well-known coroner, surgeon, and lawyer in Middlesex County.[9] Dr. Cornish's son, Francis Evans Cornish, became Mayor of London six years later. The Cornishes were among the elite of London society, and they had clout in politics, too.

Despite these successes, financial troubles were beginning to brew for Tripp and the shareholders of the International Mining and Manufacturing Company as 1855 rang out.

THE FIRST OIL WELL

Even though expansion and growth in population spawned politicians and political rallies all around him, Charles Nelson Tripp was not interested in politics. In 1856, Port Sarnia made application to the legislature to be given town status and a name change to "Sarnia". The second mayor of the town was a young Scot named Hope Mackenzie, whose brother, Alexander Mackenzie, later became the first Liberal prime minister of Canada.

Tripp could certainly have rallied strong forces to support both him and his projects if he'd been interested in getting to know the Mackenzies, but he had a one-tract mind. Asphalt to pave the roads of the world. Nothing else. Unlike Tripp, people took their politics and oil seriously in Lambton County, and they were willing to fight for both. Mackenzie was first elected to the legislative assembly of the Province of Canada in 1861. He operated a lucrative stone mason company in Sarnia. Politically, he was allied with George Brown, founder of the well-known Toronto newspaper, *The Daily Globe*, who was lobbying for, among other things, a united Canada coast to coast. At age thirty-three, Mackenzie was a man with political clout; he loved Lambton County with its silver beaches stretching for miles along Lake Huron, and its potential for river and lake trade. Mackenzie would have been the kind of man to help a friend in need, steering him back on course, but Tripp did not try to get to know him. In fact, there is little evidence that Tripp knew anyone in nearby Sarnia. He turned to Hamilton for his supplies and associates, perhaps because Hamilton was a bigger center. It was here that he met James Miller Williams when he went to the Williams and Cooper Carriage Factory to order wagons for his asphalt business in Enniskillen. The two men were opposites in every way, Williams being cautious by nature, choosing his business partners with great care, and Tripp being impulsive, accepting partners for their monetary worth alone.

In September of 1855, Tripp and his wife sold

200 acres of prime oil land to his brother, Henry, for 400 pounds. This sale seems to indicate that Tripp, who now had a wife to support, was in need of money. Almira Tripp eventually became a heavy yoke around her husband's neck.

A year after forming his company, Tripp was having financial problems. As early as October 1855, the Bank of Upper Canada obtained a judgment against him for 500 pounds, and by the end of the year, another eight people had filed notice with the Sheriff of Lambton that they wanted to be paid for services rendered or goods sold to Charles Nelson Tripp. In all, he owed 1094 pounds.[1]

From February until September of that year, he advertised in the Hamilton paper Dr. Antisell's report on his asphalt beds.[2] He was, no doubt, hoping that people would invest money in his company, but nobody invested. Tripp worked from sun-up to sun-down. He did not have an outside job that would bring in money for his living expenses, and for enormous orders — like the Paris order — he must have employed many men to help him. And he was not the kind of disciplined person to keep records of financial transactions.

By early 1856, the directors and shareholders were getting impatient. In two years there had been no profit and their confidence in Tripp's ability to do as little as break even was destroyed. In February, John B. Van Voorhis obtained a judgment against him for 1500 pounds. A week later, Tripp sold Lots 16 and 17 in the 2nd Concession — a total of 400 acres — to a group of men from Hamilton, and Williams was

among them.[3] With these two sales, he raised enough money to pay Van Voorhis but had little to spare. One judgment against Tripp from the Court of Common Pleas in Toronto asked the Sheriff of the County of Lambton to recover 83 pounds, nine shillings, and one penny for Robert Garnham, "for damages which he had sustained, as well as occasion of a certain Breach of a Certain Covenant made by the said Charles N. Tripp to the said Robert Garnham."[4]

At the same time, in Hamilton, the Williams and Cooper Carriage business suffered a blow when the Great Western Railway informed the partners that the railway company was going to make its own rolling stock. It was a good time for James Miller Williams to get involved in other ventures. In January 1856, the carriage factory sued Tripp for monies owed on the purchase of wagons.[5] Shortly after this, Williams turned away from his Hamilton business, and with Tripp helping him, began to explore for the big oil vein in Enniskillen. Tripp, no doubt, worked to pay off his debt to Williams, but there is conjecture, too, that he knew he would be leaving the area and he wanted to see his beloved asphalt beds in the hands of a trusted man.

Records show that Tripp, when on his own, worked the asphalt surface beds by boiling it in iron pots and using the thick residue as asphalt. In an unpublished study of Tripp's method of asphalt refining, R.B. Harkness conjectures that "in producing asphalt from the Enniskillen gum beds, Tripp could not help producing naphtha, as he must heat the raw product to separate it from the clay matrix.

When he had exhausted the limited surface deposit and began using the softer material and finally the heavy oil. . . he must distill off the lighter fractions in order to produce asphalt.''[6] Naphtha appears in the lower part of the gasoline group of gases and in the upper part of the kerosene group. At a later date, when pioneer oil men were doing their own refining, their stills were so crude that some naphtha and gasoline often got into the kerosene, and many a lamp and lamp lighter blew into the next world.

Dr. Gesner patented his process for extracting kerosene from crude oil in 1854. No doubt Tripp was well aware of this process and the newly-coined word ''kerosene''. In fact, since Harkness pointed out that Tripp produced naphtha, he must also have produced kerosene in his refining processes, but he probably dumped both naptha and kerosene into the nearby creek as useless by-products of asphalt.

What Tripp dumped into the creeks, James Miller Williams wanted. Williams wanted an adequate supply of crude that he could distill into kerosene and sell as lamp oil, and he felt that this bonanza lay well underground the gum beds.

The first exploratory work Tripp and Miller conducted was on the bank of the Thames River near Bothwell in Kent County, next door to Lambton.[7] It seems odd that Williams would drill in Kent County when Tripp had been working his asphalt beds successfully in Lambton. However, Williams was not interested in asphalt, and perhaps he decided to look for oil in a far-off field. As Fergus Cronin in the *Imperial Oil Review* recounts, ''they dug down to 27

Site of the first commercial oil well in North America on the grounds of the Oil Museum of Canada.

Close-up view of the first oil well showing the crude drilling and hoisting mechanisms.

feet without success, until one morning they found the hole full of oil and water. In an attempt to reach the 'pool of oil,' they tried to drive an iron pipe down the hole, but when it had gone down a 'considerable distance,' it broke, and the well was abandoned."[8]

They were discouraged with this trial run, but Tripp must have urged Williams to try again in his old stamping ground where oil had oozed to the surface to form gum beds. By the dawn of 1858, they were digging on Lot 16 Concession 2, the parcel of land that, in 1850, Murray had said was "a bed of nearly pure bitumen." Although Tripp had sold this 200 acre spread, Williams was part owner with several Hamilton men. At a later date, Williams bought out his partners and became sole owner.

It is certain that Tripp and Williams dug and searched many times in this area before they hit pay dirt. The structure of the Enniskillen oil field is eccentric in that it is built in dome formation, and accumulation of oil is directly related to the structural high.[9] At a later date, wildcatters struck it rich in one location and were often rewarded with dry holes 50 feet away.

There is one story that says Tripp and Williams were desperate for drinking water — a commodity difficult to find in this asphalt swamp — and they had dug down to 49 feet, stopping when they came to rock. They went to bed for the night, and when they returned the next morning, oil, not water, had risen to the surface in their well. Another story says the men dug several wells hoping for oil, but found only dry holes or salt water until luck led them to one particular location. They dug down, cribbing the sides of the seven by nine foot hole with logs as they went to prevent the clay from caving in. When they reached 49 feet, oil flowed freely to the surface.

It doesn't matter which story is correct: Williams had found his oil, the first oil well in North America. He bought a small steam engine to work two pumps, a force pump to take water from the bottom of the well, and (since oil always rises to the top) a lifting pump at the top. Then he purchased all the whiskey barrels he could find to store his gold. After this, he built the first commercial refinery in North America, a simple retort to process his oil for lamps.

Williams not only had the money to develop his new enterprise, but he was one of the few pioneer oil men who eventually made a fortune in this early oil patch. Tripp was not as lucky. By the end of 1858, the Sheriff of Lambton had issued fifteen judgments against him for monies owed.[10]

That year, several ads by James Flintoff, Sheriff, appeared in the *Sarnia Canadian Observer*. One read as follows: "I have seized and taken all the right title and interest of Charles N. Tripp at the suit of Andrew Stevens of, in and to Lot #13 in the 10th Concession and the southeast quarter of Lot #13 in the 11th Concession of Enniskillen containing by admeasurement 250 acres all which lands and tenements I shall offer for sale at the Court House, town of Sarnia, July 25, 1858 — 12 noon."[11] Shortly after this, Charles Nelson Tripp quietly left Enniskillen for parts unknown. "As a promoter, Tripp did his work very thoroughly," R.B. Harkness has commented. "He sought the best advice available — and he failed, principally because his operations began before railway transportation, before there was a knowledge of crude petroleum, and, before the art of refining was well-known."[12]

JAMES MILLER WILLIAMS

In 1858 a group of New York investors hired Edwin Drake to look for oil in an isolated hinterland in Pennsylvania. Drake was a jack-of-all trades. At various times in his life, he had drilled for water and worked as a conductor on railway trains, but when hired to go wildcatting, he had been unemployed for several years. In poor health, he lived in an old hotel in New Haven, and was considered a frail, thirty-eight-year-old drifter. To give himself

credentials for the new job, he sent letters ahead of his arrival to Titusville, an impoverished village of one hundred and twenty-five people near oil lands in Pennsylvania. He chose a bogus title and addressed the letters to himself — Col. Edwin L. Drake. The new identity had the desired effect. It made him an instant very important person in Titusville.

Unlike Drake, James Miller Williams had excellent credentials as a successful businessman in

James Miller Williams, the Father of the Petroleum Industry in North America.

Canada West. By 1852, he was well-established as a carriage maker in Hamilton, a growing, industrial city at the western end of Lake Ontario. In that year, the Williams and Cooper carriage factory employed forty men who not only worked on the premises, but lived there, too, in "comfortable apartments" supplied by Williams and his partner, H. G. Cooper.[1] The company manufactured various models of carriages and wagons, including railway cars for the Great Western Railway that, in 1851, reached across Ontario from Hamilton to the thriving city of Detroit, Michigan. Williams was also a popular alderman who, by the mid-1850s, had served twice on the Hamilton council. Forty years later, the *Hamilton Evening Times* published this obituary for James Miller Williams: "In 1856, he discontinued the car business and became a pioneer manufacturer of refined illuminating oil from petroleum . . . . He was not only the first to move in the manufacture of this now universal, priceless, worldwide blessing, but was actually the discoverer of petroleum itself and received medals from the British government both for the discovery . . . and for the best manufactured oils . . . and his name can now be seen on the geological records of Washington as both the discoverer of petroleum and as the pioneer in its refinement and preparation for illuminating and lubricating purposes. . . . He made the first extensive shipments of the refined article to foreign countries . . . the United States, South America, Europe and China."[2]

James Miller Williams was born in Camden, New Jersey in 1818. His parents emigrated to the United States from Wales shortly after the turn of the century. A brother, Rowland Magnus, died at age fifteen months — three weeks before Williams was born — and a sister, Elizabeth J., was born when he was fifteen years old.[3]

Williams left school early and was apprenticed to a Camden carriage maker. Reared with a love for the British way of life, he became increasingly more dissatisfied with anti-British feelings in the United States — an aftermath of the American Revolution. He wished to stay with British values and the British flag, so he studied the map of North America and decided he would re-locate in London, Upper Canada, choosing this town because it had the same name as the British city. At age twenty-two, he journeyed north with his seven-year-old sister and a forty-three-year-old retainer named Jane M. Vandroll. Here, bizarre circumstances surround Williams. John G. Taylor, a great grandson, says that he and current members of the family are puzzled because their ancestor also came north with the remains of his infant brother who'd been buried in Camden for twenty-two years. The family wonders if, in some macabre way, he wished to keep all the siblings together. There is no evidence that his parents made the move with him. The family believes that they died before he left the United States.

Two years after his arrival in Canada, Williams met a nineteen-year-old attractive brunette, Melinda Clarissa Jackson, whose family had come north from Vermont. He and Melinda were married in London. She certainly moved into a ready-made family — her

Sarnia Observer Advertiser, 5 Aug. 1858

IMPORTANT DISCOVERY
IN THE TOWNSHIP OF
ENNISKILLEN.

We lately heard of the discovery of a bituminous or oleaginous spring in the Township of Enniskillen, but our information was not of such nature as to enable us to give any proper explanation of its nature or productive qualities. We find the following in the *Woodstock Sentinel*, however, which gives some facts respecting it that have probably been obtained from the proprietor of the land who is said to live in the town or neighborhood. Perhaps some of our Enniskillen subscribers can furnish us with more particular details:

"An important discovery has just been made in the Township of Enniskillen. A short time since, a party in digging a well at the edge of a bed of Bitumen struck upon a vein of oil, which combining with the earth, forms the Bitumen. The conjecture respecting this flow of almost pure oil is, that it has its source far in the bowels of the earth — that ages have been required to form, by finding a vent, the present beds of Bitumen — and that the supply of fluid thus accidentally discovered will continue an almost inexhaustible supply of wealth, yielding at the lowest calculation, and with no greater flow than at present, not less than one thousand dollars per day of clear profit."

husband's nine-year-old sister, the middle-aged retainer, and, lurking somewhere around their first home, Williams' dead brother. There are no records to give a clue as to Melinda's thoughts and feelings on these new relatives, but the bones of Rowland Magnus remained with the family until they moved to Hamilton four years later. He was the first to be buried in the family plot in Hamilton. Elizabeth died at age twenty-one and was placed beside her brother. Vandroll remained a faithful servant until she died in 1888 at age ninety-one. She, too, is buried in the family plot.

Shortly after his arrival in Hamilton, Williams joined H. G. Cooper as co-owner of the Hamilton Coach Factory which, within the year, was re-named the Williams and Cooper Carriage Factory.

By 1852, he was the proud father of two daughters and three sons. He loved coming home at night after a hard day at work, to be with his wife and children. The long months spent on his oil lands must have been lonely for a man who enjoyed his home hearth. Oil fields, like gold fields, were the ruin of many marriages. Long absences from home, fortunes lost, quick riches, and the tinsel-like temptations of hooch and prostitutes — these managed to drive apart husbands, wives, and children — but Williams kept his family intact. Melinda never set foot on oil lands, although in later years a son, Charles J. Williams, joined his father in the refining business.

When he discovered oil in 1858, Williams was thirty-nine and appeared middle-aged, with a heavy shock of hair peppered with gray, full drooping moustache, mutton-chop whiskers cascading down either side of a clean-shaven chin. His eyes were bright and edged with wrinkles, denoting a sense of humor. One writer described him as "a careful man, simple in his tastes, retiring of disposition, practical to a degree and always dependable."[4] His granddaughter, Douglas May Williams, remembered that when she was four or five, there were often family gatherings at her grandfather's estate called Mapleside in Hamilton. Here, the old gent, in a playful mood, sometimes threw her into the lily pond.[5]

In 1858, Williams' well yielded five to one hundred barrels a day. The *Sarnia and Lambton Observer Advertiser* ran the following article on August 5th: "An important discovery has just been made in the Township of Enniskillen. A short time ago, a party in digging a well at the edge of a bed of bitumen struck upon a vein of oil. . . . The conjecture respecting this flow of almost pure oil is that it has its source far in the bowels of the earth . . . and that the supply of fluid thus accidentally discovered will continue an almost inexhaustible source of wealth yielding at the lowest calculation and with no greater flow than at present, not less than one thousand dollars per day of clear profit."[6]

By November, the Sarnia paper recorded "that the proprietor of the land in which the springs are situated, had erected a suitable building thereon, and is now manufacturing by disillation, a beautiful burning oil. . . . Its illuminating properties are so great that an ordinary-sized lamp giving a light equal to six or eight candles, can be kept burning at the rate of one quarter cent per hour."[7]

In the summer of 1859, a man on horseback visited a clearing in the Enniskillen forest and found a colony of log buildings: "The dwellings of the workmen employed [are] on three or four acres where the proprietor, Mr. Williams, has sunk a well some 30 feet deep. . . . from which he obtains by means of a pump a plentiful supply of a black, thick-looking oily substance. This, it appears, can be got in any quantity; it is then subjected to a sufficient heat to cause the finer oil to evaporate, by which means and a distilling apparatus, it is prepared for market. The waste in passing through this process was about 20 percent. When thus prepared, it is of a clear amber color, and burns in a lamp with a clear, beautiful flame, as we discovered upon trying the experiment."[8]

The operation the stranger described had certainly been there for at least two years, if not longer. It would take that much time to bring in the machinery and workmen from Sarnia or Hamilton and to build adequate accommodation for the crew. However, the oil that Williams refined was actually a foul-smelling product because of the sulphur content. In later years, Enniskillen oil was called "skunk juice" because it emitted a smell like rotten eggs. Pennsylvania oil was low in sulphur and therefore a purer product, but the oil later discovered in Ohio and other States was also tainted with the sulphur, a demon with which early distillers found it difficult to cope. An early shipment of Canadian lamp oil was stored at dockside in Montreal next to a load of flour and bacon, saturating the food with its vile odor. The food merchant sued the oil company and won — and the company went bankrupt.

The smell of rotten eggs did not, however, keep speculators away from Enniskillen once word was out that oil had been found in abundance. The boom was on. A shack and tent town began to grow and was called "Oil Springs." By August of 1858, the Great Western Railway was in operation from Toronto to Hamilton to London and Wyoming, the latter a railroad stopping point 15 miles north of the oil fields. Wyoming grew quickly into a well-laid out village with a blacksmith shop, hotel, saw mill, grist mill, and lumber yard. Williams hired teamsters to take his oil in barrels from Oil Springs to the railhead in Wyoming, but this short stretch was described by a roving reporter as "one long mud hole in which the horses sometimes threaten to disappear. Two barrels on a stoneboat are all that a team can take."[9]

In 1858, A. C. Ferris, a New York entrepreneur, ordered 180 barrels of lamp oil from Williams, and it was duly delivered that summer. Williams was also selling both crude and refined oil in Toronto, Hamilton, London, and Sarnia — and was planning to extend his market to England. By 1859, he owned 800 acres around Oil Springs. The assessment rolls show that there were 166 tax payers, and double that number of squatters and lessees, all speculators who had come with axes, shovels, divining rods, hopes and dreams.[10]

Oil men have an uncanny way of knowing where and when there is a big oil find. News of this gold travels quickly, even if the bonanza is hidden in a remote corner of the world. Edwin L. Drake, struggling to find oil in 1859 in Titusville, must have known about Williams' success the year before in Canada. And, no doubt, Williams was aware of Drake, too. After drilling 69 feet on 27 August 1859, Drake found oil, but this was not a spectacular well. It gave nine to thirty-two barrels a day for a four-year period, and Drake was happy with this. He did not drill more wells, nor did he buy up large tracts of land, as Williams did. He liked to spend long hours fishing in nearby lakes and rivers, or playing cards in taverns. He did not foresee a world-wide market for oil. By 1863, he had acquired $15,000 in the oil region of Titusville, and three years later he was ill, in need of constant care, and broke. In 1873 the State of Pennsylvania gave him a small, life pension for the part he played in finding oil. Other than that, he received no recognition for his work.

Williams, though, quickly bought up large tracts of land in Oil Springs, drilled more wells, and became president of an oil company that did big business in North America and abroad. Although he won honors in England for being the first to produce and the first to refine crude oil in North America, Williams did not receive any rewards from the Canadian government for his involvement in launching the petroleum industry, only the prosperity and notoriety that came with being a successful businessman.

JIGGING DOWN

As Charles Nelson Tripp had done before him, James Miller Williams quickly formed an oil company that he called "J. M. Williams & Company." He sold his carriage business and went into oil production and refining wholeheartedly. Ads for his new enterprise began to appear in newspapers in Hamilton, Toronto, Sarnia: "OILS! OILS! OILS! Illuminating & Machinery Oils at prices much below all other coal oils — illuminating oil, 70¢ a gallon, machinery oil 60¢ a gallon; crude oil, 1000 – 4000 gals, 25¢ per gallon; l000 – 100,000 gals, 16¢ per gallon." [1] Williams also sold lamps. A staunch Presbyterian, he is credited with donating twenty-four lamps and the oil to keep them burning *ad infinitum* to the Wentworth Mission, the forerunner of St. Andrew's Presbyterian Church, Hamilton. [2]

Williams can also be credited with introducing the "spring-pole" method of well drilling to the industry. The spring-pole method of drilling was also called "jigging down" because the operators looked as if they were doing a jig, the popular dance of the time. David and Joseph Ruffner, who dug wells in West Virginia to obtain salt, became dissatisfied with conventional digging methods as early as 1806. As Max W. Ball notes in his book *This Fascinating Oil Business*, the Ruffner brothers "attached a 2 1/2 inch steel chisel bit to a long iron drill and raised and dropped it by means of a rope and spring-pole." [3] Although primitive by modern standards,

J.M. Williams & Co. advertisement in The Hamilton Spectator & Journal of Commerce.

A spring-pole drilling rig (also called the "kicker" method of drilling) and crew at Oil Springs in the 1860s.

this method of drilling was faster and more efficient than digging. In a nutshell, it consisted of a spring-like pole from which the drill or chisel was suspended over the well and the pole balanced on some sort of support. On the end of this pole there was a stirrup where a man put his foot and danced up and down. John S. Ewing in his unpublished history of Imperial Oil explains that the "action of this jumping gave the impetus which lifted the drill and allowed it to fall in a series of sharp jerks."[4] This procedure was also called "kicking down" because sometimes it looked as if the men were kicking the pole.

A world traveler spent a few days in the Enniskillen oil field in 1860, and described the scene: "In those wild-wild woods there were hundreds of men all quiet. . . so quiet. . . intent on their work, and the only sound was the thump, thump, thump of the spring-poles."[5]

This method of drilling was, however, not new. In Asia, the Chinese had used rope-boring in a similar way several hundred years before. But the Ruffner brothers introduced it to North America. Williams had used spring-pole drilling near Bothwell in 1856, and when the pipe broke in the hole he had become discouraged and had abandoned the well. Now, three years later, he decided to try again, and this time he was successful. The Toronto *Daily Globe* in 1861 reported that well No. 27 on Williams' holdings was "sunk 46 feet to rock; bored 100 feet in rock. This well averages the large quantity of 60 barrels a day. . . . It has been in operation for two years."[6]

In 1861 Williams re-organized the J. M. Williams Company and called his new enterprise the Canadian Oil Company, with capital set at $42,000 and five shareholders: Williams himself, John Fisher, W. P. Fisher, Isaac Jameson, and Nathaniel D. Fisher. Williams held the controlling interest at $14,000.[7] This company endured and grew for thirty years, until Williams' death in 1890. The charter authorized the company to mine, manufacture, and sell oil with the principal places of business being Hamilton and the site of operation, Enniskillen.[8]

By this time, Williams had five wells in operation, yielding from 600 to 800 barrels a day. Other wildcatters all around him had punched another 400 wells out of the terrain.[9] Men in knee-high boots sloshed through a muddy, oily ooze of clay. And the sheriff was busy re-possessing land dumped by partners who had spent their life's savings on dry holes. An Irishman purchased 100 acres from Williams, and, after sinking a well that yielded only salt water, he found himself destitute. He owed Williams $300 for the land, and Williams took him to court. The sheriff seized the land and offered it for sale at the court house in Sarnia where an American purchased it for the grand sum of 35 barrels of crude.[10]

William Edmund Logan, in spite of his expertise in geology, did not invest any capital in the oil region, but it is certain that he visited the area many times, for he often gave updated information to the government on activities there. He reported that prior to 1861 there were 5,529 barrels of oil shipped from Wyoming, by railroad, to customers abroad.[11]

Alexander Murray, whose report in 1850 alerted Logan to the presence of oil and asphalt in Enniskillen, was employed by Charles Nelson Tripp, probably for survey jobs. In 1856, Murray sued Tripp for monies owed to him, but, no doubt, he never collected, like many other people who were left with outstretched hands when Tripp skipped town.

But the biggest surprise of this period was the return of Henry Tripp to Enniskillen. At age thirty-two, unmarried, Henry had been working as a plate photographer in Petersburg, Virginia, and he was not interested in wildcatting. It appears as if his chief aim was to protect his brother's lands and to do some land speculating on his own. He must, however, have been communicating with Charles Nelson, or he wouldn't have known that the sheriff was chasing him and his lands were in danger of being re-possessed for back taxes.

The original lot where Charles Nelson Tripp worked his gum beds and Williams found oil had an interesting history. In 1858, the 200-acre spread was purchased by Williams and a group of Hamilton men, and later that year, Williams bought out his partners for $1,000. In 1860, Henry Tripp acquired this land and re-sold it to a Sarnia physician for $500. A year later Williams bought it back for 100 pounds, and a few months after this the land was subdivided and he sold one small lot for $20,000.[12]

In all of Henry's land dealings, none is more of a mystery than a 100 acre lot he purchased in June 1860. At that time, Almira Tripp, the wife of Charles

Nelson, was boarding in London at the home of a Mr. R. Galbraith, a bookkeeper who lived on the premises with his wife Mary, six-year-old son Ric, and one-year-old daughter Beatrice. On census records, Almira marked that she was "married," then she scratched this out and marked "widow."[13] On 4 December 1860, Galbraith wrote a letter to the crown land agent in Quebec:

Sir: Would you be kind enough to inform me at as early date as possible if Lot. No. 13 in the 11th Cons. Township of Enniskillen (East half) was ever patented or located to Charles N. Tripp — Or if there are any other Lots in the same Township or Townships patented or located to him — As he left hear some four years ago and has not been heard of since, his wife who is now hear and haveing no means of support wishes to find out if there are any of these Lands she can dispose of as he left no memorandum of his lands after him. Hopeing you will be kind enough to give a List of lands Patented or Located by him in the different Townships, also the amt due on each Lot.
Yours Respectfully
R. Galbraith
for Mrs. C. N. Tripp [14]

There is no record of a reply from the crown land agent, but the land mentioned (Lot 13, Concession 11) was purchased by Tripp in 1853, before he was married, and, in 1856, a year after his marriage, the sheriff seized the 50 acre lot and re-sold it.

In June 1860, Henry purchased 100 acres in the Second Concession — land his brother had never owned. Six months later Almira sold that land for $1,000. There is no evidence that Henry transferred the title into Almira Tripp's name. He was, however, a man who had great empathy for people. All his life, he took care of two older, unmarried sisters, and he constantly picked up the pieces of his brother's shattered dreams in Enniskillen. It seems very likely that he sold the land to Almira for next to nothing because he felt sorry for the way his brother had treated her.

Henry did not make any money from land deals in Enniskillen. In fact, he lost money. He did not have the luck or speculative talent that Williams seemed to have, nor did he have the ready cash for investment. He did manage to save three of his brother's lots for the next eight years, no doubt covering the taxes to keep the government from snatching them away.

In 1865, Henry was back in his home town, Schenectady, where he became much loved for his historical pictures and well-known as a "Pioneer Lensman," as Larry Hart, a reporter, called him, noting that photographers of that era "were content to work in their skylighted studios and photograph posed individuals or families. But not Henry. He rigged up a platform cart on which he placed a small tent that could be closed tightly. This dark room was pulled on a light wagon by a horse. His camera was a cumbersome one, with extending bellows and large, solidly-made plateholders. After parking his wagon, he made a careful study of the view to be taken and then set up his tripod. With this preliminary phase of his work out of the way, he then darted into his dark room, poured the photographic emulsion over the clean glass plate, placed the plate into the holder and, while the emulsion was still wet, rushed with his camera to the scene to be photographed. Adjusting his lens, he made the exposure. Another dash to the dark room followed. There he had to develop his plate, 'fix' it and put it into a 'bath' before the emulsion ran. The printing process took place at his studio, sometimes many miles away from the view he had just captured. Henry Tripp was a creative artist capable of catching the last bit of charm in an historical scene and transferring it hurriedly to a wet glass plate," Larry Hart concludes. [15]

As far as is known, Henry did not try to capture the "last bit of charm" in the oil fields of Enniskillen. If he did, no pictures remain as mementos, unfortunately. Still, other colorful oil men would leave their mark on the Enniskillen landscape, hard oilers like Hugh Nixon Shaw, the evangelical oil man.

HUGH NIXON SHAW

In late December 1860, military forces at South Carolina laid siege to the federal garrison at Fort Sumter marking the beginning of the American Civil War. There was still an anti-British feeling in the United States, a hangover from the Revolution that had severed British ties more than half a century earlier. Many Yankees felt that Britain should be punished because, when the Civil War began, she showed a sympathy to the cause of Southern Rebels; a few people actually wanted American armies to fight for and conquer the British territories to the north.

The American Civil War made many Canadians feel threatened; but, in the oil fields of Enniskillen, the Civil War was not a menace at all. The Southern States were the sole manufacturers of turpentine made from oleoresins of coniferous trees from which the cheap, dangerous lamp oil camphene was derived. Despite its volatile nature, camphene was popular, especially in cities and towns where it could be easily purchased. With the outbreak of war camphene was no longer readily available, and people began to clamber for kerosene, thus stimulating the oil business.

Oil men, unlike other people in Canada West, did not fear an invasion by the United States, even though they were only 27 miles from the undefended border, probably because there was a rush of Americans to Oil Springs just before the war broke out. The American preserve in Oil Springs was so dominant that by 1861 the newly-laid out town ignored Queen Victoria's Birthday — the 24th of May — and celebrated American Independence

Day, the 4th of July, with parades, games and races.[1] At this time, the main road of the town was springing to life with hotels, groggeries, blacksmith shops, post office, a town constable, and a small holding unit for law breakers. Beyond this, a forest of ash-pole derricks greeted visitors.

Enniskillen was a haven for the hardy black ash tree that thrives in rich, moist, swampy lands. The wood is heavy, hard, tough, and does not splinter. Oil men soon learned the value of this wood, and very quickly they depleted the area of black ash. The trees were especially good for the construction of three-pole derricks that sat 55 feet above a well like the skeleton of a tepee. These derricks were used to pull pipes and do repair work on the wells. The three poles were bolted together at the top. A pulley was placed on a U-shaped iron that hung from the bolt so that workers could let a rope down to pull up pipes.

Tragedies began to happen in this sea of ash-pole derricks, as reported in the *Toronto Leader:*

A fatal accident occurred in the Township of Enniskillen. Three brothers by name of Cook had sunk a well down to the rock without finding oil and then commenced to drill into the rock. After going some thirty feet, they struck some cavity containing gas, which burst forth and became ignited by the light of a lamp which stood upon a platform placed some ten feet below the surface, and it produced an explosion which was heard at two or three miles distance. Two men were down upon the platform at the time; a third was just gone up, and was in the act of stepping on the surface. One of those on the platform — one of the brothers Cook — was killed; the other so seriously injured that he is not expected to recover; while the one who had just reached the ground was so stunned that he lay for some time in a state of insensibility. It thus appears that oil digging is attended with dangers not hitherto dreamed of.[2]

Into this raw, rough, dangerous territory came a gentle Wesleyan Methodist lay preacher who turned Oil Springs into the liveliest, gaudiest town in North America.

Hugh Nixon Shaw was born near Dublin, Ireland in 1812. His mother died at his birth, and he was adopted by an uncle — William Nixon — who was a titled, wealthy landowner. Shaw was destined to inherit the estate from his uncle who had no children of his own, "but he was too restless to be shackled to the management of farm lands," as Ben Fiber has noted in the *Sarnia Observer*.[3] At age twenty-one, he booked passage for Canada.

The stocky, clean-shaven young man quickly joined a community of Irish immigrants near Toronto where Bartholomew Bull, a charismatic Methodist lay preacher, welcomed newcomers like a father would welcome lost children. Blue-eyed, tall, good-looking, and well-educated, Bull with his wife Elizabeth had come from Tipperary, Ireland, landing in Canada in 1818. Both had been converted to the Wesleyan Methodist faith by the brilliant young Irish missionary, Gideon Ouseley.

Bull and Elizabeth settled on a 200-acre farmstead in York Township, and here, in their first home, a log house, they raised five sons and four daughters. Here, too, he made provision, in the words of the church historian William Perkins Bull, "for the livelihood and welfare of the families of immigrants, otherwise friendless and helpless, whom he had gathered together for the clearing and developing of his own lands."[4] Hugh Nixon Shaw was one of those befriended immigrants. Although shortly after his arrival in Canada, he went to work in Goderich — a town with a natural harbor on Lake Huron 80 miles north of Sarnia — Shaw often came back to the Bull farmstead like a child returning home.

Eventually, Bull built a beautiful, two-story brick house that he called "Springmount," the first brick dwelling in York Township. This home was often the site of Methodist meetings when prayer and praise shook the rafters.

Anne Bull was only fifteen years of age when Shaw first came to the farmstead, but this middle child of Bartholomew's soon grew to love the gentle young man who spoke with an appealing, lilting Dublin accent. In 1833, he converted to Methodism, and two years later he and Anne were married. By this time, he had become a fervent convert, trained by Bull in preaching, prayers, and the Bible. Although zealous in his faith, Bull was no retiring monk. He liked people. The Bull home was often full of laughter and friends, and he pushed his children to higher education. Later, his sons became doctors, lawyers, and politicians, and he, himself, was a

Hugh Nixon Shaw drilled the first wild well in Canada.

friend of Egerton Ryerson who founded Upper Canada College and became the first principal of Victoria College, University of Toronto.

Although life was happy and the future looked promising for Shaw and Anne, several years later they would know tragedy in the oil fields of Enniskillen. After their marriage, the couple lived in Goderich where Shaw painted and decorated the hulls of ships. By 1842, seven years later, they owned two houses in Goderich, a frame one-story dwelling and a two-story brick house with fireplace.[5] They were certainly not poor by the standards of that era, when the average family rented (sometimes owned) a single, small house made of logs. After ten years in Goderich, Shaw tired of working on ships and he took Anne and their three children back to the Bull home. The following year he went into business for himself as a storekeeper in Cooksville, a neat, trim village just south of the center of Toronto Township. Here, Anne gave birth to two more children, and a large family to feed meant that Shaw had to work hard, long hours. On top of tending the store, he took on the added task of being an agent for the Equitable Fire and International Life Assurance Company.

In 1849, there was a stampede to the South Fork of the American River at Coloma, California where gold had been found. The following year, Shaw joined the stampede, but there is no evidence that he struck it rich. By 1851, he was back in Cooksville. Nine years later he was still a storekeeper and insurance agent, and as such, he must have heard about the oil discoveries in Enniskillen.[6] Instead of heading

for the oil field, he made a trip back to Ireland. Here, he learned that his uncle had disinherited him and had bequeathed his lands and worldly wealth to his two older sisters. This news did not worry him, for his whole life was wrapped up in his family and his religion. Before leaving Ireland, Methodist friends presented him with a silver tea service engraved with the words: "To Hugh Nixon Shaw, Esq., as a token of respect and esteem on his return to Canada West. Dublin, June 1860."[7] Back in Cooksville, the urge to go to the boom in Oil Springs surfaced. He left his oldest son, Bartholomew, age twenty-four, in charge of the store, and journeyed to Enniskillen.

Wandering among the ash-pole derricks and spring-pole drilling rigs, Shaw liked what he saw. The idea of instant wealth, if one were to hit pay dirt, appealed to him. Also, as a staunch Wesleyan Methodist who had often preached at the Bull homestead, he perhaps thought that this oil field was a sea of lost souls needing conversion. Whatever the appeal, he decided to stay and he allied himself with three partners to drill a well.

Sixteen hundred men were now scrambling for a fortune in this swamp land. Many of them were a tough breed who chewed tobacco, spat, swore, and swilled hooch with gusto. And there were shysters among them who would steal gold from their mothers' teeth without any remorse. In reality, it was alien turf for the gentle lay preacher. At a later date, an author who called himself "J.S.", writing in the *Christian Guardian*, said of Shaw, "His mildness and affection made [his] home happy. He was,

Three-pole ash derricks stretch across the land at Oil Springs in the 1860s.

though intelligent, too unsuspicious for designing men, by whose nefariousness he suffered much."[8]

Within the year, Shaw and his partners drilled a dry hole. By this time, there were, along the main street, several hotels with high-sounding names: Michigan Exchange, Twin Sisters, New York House, Royal George. A *Toronto Leader* reporter, visiting in September 1861, described this scene: "The Michigan Exchange. where your correspondent lodged, was a large frame building, built in the barn style, offering free ventilation. About fifty of us slept in the upper room in beds arranged like ships' berths, one above another, and two in a bed at that. Many slept on the floor with only a counterpane under them and their boots for pillows. Many were turned away from the door." The same reporter commented on the oil fields: "I saw wells yield day after day, from ten to fifty barrels. . . . There will be double 300 wells before a year at the present rate of sinking. . . . One Toronto firm is buying some 3000 barrels [of kerosene] whilst another is introducing it largely into Europe. Once the European market is opened, it must create an immense trade for a substance from which is made refined toilet soaps, candles superior to wax, oil that gives a cheaper and not less beautiful light than gas . . . and the best machine and lubricating oils yet known."[9]

The hotels were wild, gaudy places sporting saloons and burlesque shows. People honored the Sabbath, however, and the hotels were then ghost-quiet, with the exception of one where church services were held in the saloon. It is very probable that Shaw conducted these services in fervent, Wesleyan Methodist style. William Perkins Bull has said of the itinerant Methodists that "they counted their conversions and often cared very little to what lengths they went in securing new members. They were not politicians, but apostles. If there was a lack of restraint in their preaching, it was because they were more concerned about souls than about syntax, more interested in people than in parish boundaries. . . . One exhorter is described as leaning over the rough pulpit in a burst of sympathy and sorrow, crying, 'Oh, ye hell-bound souls, turn or burn!'"[10]

Although kind and gentle, Shaw belonged to this passionate, zealous class of Methodist, and it is not surprising that within the year his partners quarreled with him. Perhaps he preached to them on the job, or perhaps they did not like him because he would not carouse with them in the groggeries. In any case, they dissolved the partnership, and alone, Shaw left.

William E. Sanborn, an American entrepreneur (also listed as an Enniskillen gentleman) owned land and was doing exploratory work for oil. Shaw leased from him one acre in Subdivision 10, Range B, Lot 18, Concession 2.[11] The lease bound him to give one-third of his oil revenue to Sanborn, *if* he hit pay dirt.[12]

It was spring 1861. Shaw was low on money after his failure of the previous year, and with a working capital of only $50, he set up his spring-pole drilling rig and began the arduous task of "jigging down" to find oil.

JOHN HENRY FAIRBANK

Although Shaw did not fit the mold of a rough, tough entrepreneur in a rough, tough boom town, he was tenacious and stubborn, two traits that would be of great benefit to him in the months that lay ahead. Another man who was tenacious and stubborn arrived in Oil Springs about the time that Shaw began to drill for himself. Although different in temperament, these two men would become friends, and individually they would make a considerable impact on the oil industry.

John Henry Fairbank was a slight, bearded young man who lived in Niagara Falls, Ontario. He came to Enniskillen to survey a parcel of land for a client named Julia Macklem. When he reached London, he heard about the oil boom, but on arrival in Oil Springs he was not prepared for the sight of derricks reaching to the far horizon, and a mass of humanity scrambling for riches. He was so mesmerized by the fever pitch of excitement that he had a difficult time keeping his mind on the humdrum survey job.

Fairbank was born in Rouse's Point, New York in 1831. As his biographer Edward Phelps notes, Fairbank "came from a long line of New England colonists and soldiers. The first of these forbears, Jonathan Fairbank, emigrated from Yorkshire, England in 1633 with his wife and six children."[1] Fairbank's father — Asa Fairbank — farmed near Rouse's Point where he also worked as a deputy sheriff for Clinton County. Asa died when Fairbank was eleven years old, and, since the boy was an only child, he became very close to his mother, Mary Oliver Fairbank. In his own words, he explained his early years:

I received some education at the Champlain Academy, and more hoeing corn and driving oxen. I seemed adapted to oxen. My boyhood pride was in a yoke of white steers which I had trained from the time when, as weaned calves, they snorted milk up my shirt sleeves, until they could make the fastest time on the steer record.

There are few mistakes open to boys or men that I have failed to discover. If I had any brother or sister I think we would have found them all.

At the age of twenty-one I came to Canada and followed surveying. I admired Niagara Falls, also one of its daughters, which annexed me to Canada.[2]

In the 1850s, Fairbank traveled around Canada West on surveying jobs, and when returning to Niagara, he boarded at the farm home of Hermanus Crysler. Here, he met Crysler's son, Abner, and his daughter, Edna. The Cryslers were descendants of well-known and highly-respected Loyalist families, but, in spite of this, they warmed up to the penniless immigrant from the United States. In September of 1855, Fairbank and Edna were married, and the Cryslers welcomed him into the family.

John Henry Fairbank was another founding father of the Canadian petroleum industry.

The couple's first son, Henry Addington, was born a year later, and in 1858, a second son, Charles Oliver, arrived. Shortly after this, the couple purchased a farm near Niagara Falls. By the end of 1860, however, Fairbank realized that farming and surveying were not providing enough income for the living expenses of his growing family. After Charles was born, he had to borrow $300 from Abner to make ends meet.[3]

He loved farming, but he wanted a decent living, too. When he stumbled upon Oil Springs, it must have seemed like the saving grace in his life. He was typical of the penniless speculator who would risk the family farm and all he possessed in the hope of hitting pay dirt. Late in life, remembering his first encounter with Enniskillen, he said: "Somehow I had faith there was something in this oil for me — I never lost faith."[4]

By July 1861, he finished the survey job and borrowed $500 from his wife's father. After paying off current debts, he had $351 left, and he put $30 down on a half-acre piece of land that he purchased for $300 from James Miller Williams. He paid out $150 to begin digging a well, and he hoped to spend another $100 on this project within a few days.[5]

These early oil men dug a hole roughly six by six feet square and cribbed it with logs to a depth of 40 to 60 feet. Then they set up their spring-pole rig and began to drill. Fairbank worked all that summer and fall on the digging because this had to be done before the ground froze. He wrote to Edna, lamenting the risks of drilling, and apologizing for his obsession with oil:

So you see that if I went on with drilling imme-diately, I should have to risk everything in that without prospects of immediate returns should I get oil and as drilling uses up money awfully fast, I think I will wait a little as I lose nothing by so doing, and may do pretty well with my oil specula-tions. Well, I have got back to oil again in full blast — really I must have it on the brain, the next time I want to write you I can just pour a little in an enve-lope and send you.[6]

Edna hated the oil business, and she despised the shack and shanty town that had grown up around the mud and spring-pole drilling rigs. She was definite-ly not a pioneer, and she made this known to her hus-band. Also, she complained of overwork and ill health. With the help of five hired men, she managed the family farm that would eventually be heavily mortgaged to support Fairbank in his oil ventures. Her father, too, mortgaged his farm to raise the nec-essary money to bail Fairbank out of debt. She had a right to be worried about finances, but her constant pessimism with regard to Fairbank's endeavors in the oil patch must have rubbed him the wrong way on occasion. She constantly turned down his pleas to come to Oil Springs, although at a later date she did make the journey, but not to her liking. There is a rumor, probably correct, that at this difficult time, Fairbank scooped up waste oil that ran freely in creeks and sold it between Wyoming and London as machinery oil. In a letter to Edna, he spoke about himself and a partner, J. H. Eakins, taking ten barrels of crude oil on the road. They cleared $8 to $10 each, and he said when customers want more oil they can order by mail, thus cutting out traveling expenses.[7]

He clung tenaciously to his marriage *and* the oil business, although enormous difficulties and financial problems plagued him. He said in a letter: "My well is steadily going down. I expect to reach the rock at a depth of 50 or 55 feet, which we shall accomplish this week, having done this, and got tube in all ready for drilling." In the same letter, he spoke about health problems: "Not feeling well; working hard; considerably troubled with pains in the bowels, a common complaint which some attribute to the water supply."[8]

Death from accidents and illnesses constantly stalked these oil pioneers. In 1858, a report in the Hamilton *Daily Spectator* claimed that tuberculosis was the biggest killer that year, with bowel complaints a close second, infant diseases third, and measles fourth.[9] Syphillis was on the rampage, ready to pounce on the sexually active, and cholera epidemics some-times swept the nation. Fairbank, in a letter to Edna from Preston, Ontario, described the plight of the immigrant in the cholera epidemic of 1854:

I think you and yours have thus far escaped this singularly fatal epidemic which seems to be more prevalent at the Falls than any other place I hear from. Preston is not entirely free from its pres-ence but here as elsewhere it is chiefly confined to those unfortunate emigrants who seem doomed to encounter all the ills to which man is liable.

Having bid farewell to the homes of their child-hood to which the Europeans' affections cling with a tenacity like that which binds the heart of the red man to the graves of his fathers, having undergone the privations and sufferings often of extreme poverty and survived the casualties of emigration, they are met in the land where they had fondly expected to find a free and happy home for them-selves and their posterity, by a fell disease in whose embrace in a few hours they wither. A stranger's hand consigns their bones to rest, not in their fatherland, but to a stranger's sepulchre, and they are forgotten.[10]

A descendant of one of the early oil men said the "Black Plague" killed many oil workers in the boom years at Oil Springs. "At one time, coffins were piled up — one on top of another," he lamented.[11]

Patent medicines cashed in on all maladies, making a fortune from people's pain, and there are no credible sources that say they worked. "Radway's Ready Relief" was one such medicine. A single column ad, eight inches long, regularly ran in the Hamilton *Weekly Spectator*. It said this medicine would definitely cure measles, small pox, pleurisy, typhus fever, scarlet fever, leprosy, syphillis, cholera, hysteria, and bilious attacks. "One application exter-nally or a few drops taken internally will free the suf-ferer from the most violent and terrible pains and restore the weak, feeble and prostrated frame to strength and vigor. Price — 25¢ a box. Sold by drug-gists and dealers everywhere."[12]

Fairbank often complained of the cold, the heat, the mosquitoes, the rain, mud, and bowel problems. He remained in Oil Springs in spite of the hardships, drilled his well, and, early in 1862, he struck oil. This first well, which he called "Old Fairbank," was often temperamental, yielding oil some of the time, and going dry some of the time. Two years later, he recorded in his diary that "the old well — good boy — has done big this twenty-four hours, ending today at noon, some 45 barrels. Net profit of day $150, a big day's work, the biggest ever made by me or probably that I ever shall make."[13]

Fairbank wanted desperately to be a producer, and, although he didn't like refining, he was a refin-er by necessity. When "Old Fairbank" came in, he suddenly realized that there were new difficulties and expenses he had not dreamed of. He could not find barrels in which to store his oil, and when he did find them, they all leaked.[14] The railway station was 15 muddy miles north. A Toronto *Globe* reporter said of these miles: "The number of mud holes is something wonderful. . . . The holes bury horses to their shoulders and retain wagons firmly in their grasp. Wagons pushed aside into the bush or still sticking in the mud, and piles of lumber on the road, tell where attempts to reach Wyoming or Oil Springs have been abandoned in despair."[15] And when they finally reached the railhead, what then? There wasn't much use sending oil anywhere unless there were people at the other end who wanted it. These men had to do their own marketing, and since lamp oil brought in the biggest revenue, they had to do their own refining. All of this took money, and they had to spend hard cash long before they reaped any rewards. In his book *This Fascinating Oil Business*, Max W. Ball recorded one oil man's poetic response to these trials and tribulations:

> The rockhound reeks with his lore profound
> Of domes and saltcellars underground,
> Mysterious depths pretends to see —
> Then guesses at where some oil may be.
>
> The promoter may condescend to sell
> Some gilt-edged stock in his drilling well.
> Will the well blow up and sheriff frown,
> Or the shares go up as the well goes down?
>
> The skillful stillman is seldom still;
> Much depends on the stillman's skill;
> If he opens a valve that he ought to close,
> Up with the still the stillman goes.[16]

While John Henry Fairbank was struggling with production and refining problems, Hugh Nixon Shaw was facing drilling and credit problems a few hundred feet away.

Well commenced 27 July 1863; oil struck 3 October 1863

Log:

		Total Depth
Surface (conductor)		65 feet
Hard rock to first oil	75 feet	140
Second oil about	15	155
Main vein	18	173
More gas at	2	175
Hard rock to 4th gas & oil abt.	18	203
Soft rock	26	229

New well had steam pump installed at end of first month.
The production paid for the well before the end of the year

Cost:

Digging surface	$ 60.00
Drilling 154 feet	284.00
Conductor [wood?]	9.00
Derrick and fixtures	3.00
Piping	10.92
Pump	93.28
Tank	15.50
One week pumping and fitting on pump	15.00
TOTAL	$490.70

Log and cost for drilling a well, from John Henry Fairbank's Diary, 1862–64.

WILD WELLS

As the summer of 1861 waned and the oaks, elms, and maples splashed rainbow colors across the land, Hugh Nixon Shaw became a joke in Oil Springs. The deepest well was one of Williams' at 65 feet; by late fall, Shaw had far surpassed this depth at 150 feet.

That year, a name unique to Oil Springs was used to describe veteran drillers like Williams. These men were called "hard oilers." Newcomers invented the name when they realized that, in dry weather, the clay soil became hard as granite. "In order to bust through that kind o' crust and find oil — ya gotta be a hard oiler," was their motto. The name stuck like damp clay stuck to boots. In fact, when Canadian drillers went to Europe and Asia at a later date, the name became a kind of code. On crowded buses, in railroad stations, it was common to hear a shout, "Hard Oilers," or "Hard Oils," and if there were any in the crowd — and often there were — they found the voice.

These "hard oilers" laughed at Shaw. They told him that oil did not exist at that level; he might as well back off and try another location; he had hit a dry hole; he was crazy to be drilling that deep. But Hugh Nixon Shaw was stubborn — or perhaps he was sure that the Great Man Above was looking over his shoulder. There is a rumor that, at this time, he told a friend that he'd had a dream. "I was walking the streets of the New Jerusalem, and the rock poured me out rivers of oil," he is reported to have said. Perhaps this dream kept him going in spite of the escalating, scathing laughter of hard oilers, and in spite of debts that had been piling up for months.

How much money he owed and the names of the Good Samaritans who gave him credit have all been erased through the years, but it is very probable that he borrowed from his wife's father, the famous lay preacher Bartholomew Bull. Bull had become a well-known, highly visible business man in Toronto, selling lumber off his land to the city. He supplied some of the square timbers for the Queen's Park parliament buildings, and accepted chopping, logging, and building contracts from government officials.[1]

In the winter of 1861, a reporter from the *London Free Press* put Shaw's plight into words: "He commenced operations in Oil Springs in June last. Onward he pushed his work till he excavated 45 feet of soil, then struck the rock and bored 100 feet; then reached 155 feet without oil, became greatly discouraged, and had heroically to battle against despair. Means were exhausted; hope almost extinguished; credit gone; he was on the eve of utter despair."[2]

Snow filled the landscape and ice lay thick on the creeks. Christmas was bleak; Shaw spent the festive season kicking down his spring-pole to send the steel drill bit deeper into the bowels of the earth. He had employed two helpers — Hugh Smiley, who often relieved him on the kick board, and Jack Coryell, who turned the tools. By the time that January rolled around, bringing in the new year, all three were depressed that they had not hit oil. When the well needed more casing, supplied by a local tinsmith, Smiley and Corywell purchased it, knowing that Shaw's purse was empty again.

There is every indication that Anne Shaw stayed on in Cooksville, helping to run the store, but, like Edna Fairbank, she must have hated this newly-invented substance called kerosene that threatened to put her family in the poor house. And yet Anne's religious convictions were as strong as her husband's. No doubt she believed that God wanted him in Oil Springs, and He, then, would take care of him. In spite of being surrounded by her children, she still missed this adventuresome husband she loved with a great passion.

On 15 January 1861, Shaw told Smiley and Coryell that he would drill one more day, then shut down all operations. Michael O'Meara, in his article "Oil Springs: The Birthplace of the Oil Industry in North America," explains what happened: "Shaw went back to his rig the next morning, discouraged and beaten, an old man with one spark of hope left, and began drilling again. When he had chopped one foot deeper, a loud crack sounded from the bottom of his well that could be heard over the entire field, and moments later heavy, thick oil shot up to tree-top level, splintering his drilling rig."[3]

If hard oilers had been hit with lightning, they

The site of the wild wells at Oil Springs in the 1860s.

would not have been as shocked as they were when this gusher exploded on the scene. They were laughing no longer. Shaw, who was a nobody before, became a somebody within a few hours.

A Toronto *Globe* reporter visited the scene a month later, and wrote:

As we descended from the high land into the flats, our attention was caught by the admonition that "No Smoking Allowed," then we saw a heterogeneous mass in which were men, women, children, sleighs loading and unloading barrels empty and barrels full, barrels clean from the cooperage and barrels smeared from the well. Amid hundreds of spectators, some men were making bung-holes in the new barrels and others were engaged in filling them, and still others were clearing the passages with every variety of noise and vociferation, while the wondering spectators were standing from one to six inches in a black, greasy matter, the mere waste of this extraordinary well, which in making its way to the creek, covered the surface of the ground for many rods.[4]

Fortune seekers flocked to the scene. Reporters from Montreal, New York, Detroit, and Pennsylvania rushed to interview Shaw. When they asked him what had happened, he quietly quoted Job 14:6, "And the rock poured me out rivers of oil."

His dream had come true. Verily, he was witnessing the gold of the New Jerusalem. The well — the first wild well in Canada — spewed oil into the heavens at 2000 barrels a day. Shaw did not know how to control it, and even if he had known, he had not purchased any barrels in which to store it. He was offered $25,000 in gold to sell the well, but he declined to sell.

Although Williams was the first in North America to refine oil from his own well and sell it commercially, Shaw's gusher was not the first wild well to explode on this continent. Ten months before in April 1861 drillers hit a gusher in the oil regions of Pennsylvania, close to Titusville. Three thousand barrels a day spewed into the heavens, and gases ignited causing an explosion and an inferno that killed nineteen people.

Shaw was lucky. He lost only his light drilling tools, although he did not know that tragedy lay ahead.

A week went by with oil raining from the sky, putting three feet of black liquid into the creeks and laying it, like thick molasses, on the ground. Then, oil men who'd had experience in the Titusville area of Pennsylvania helped Shaw insert a four-inch pipe into the well; into this they put a three-quarter inch pipe. The vacuum between the two was filled with leather packers, seven or eight inches deep, that were filled with flax seed. The moisture swelled the packers to fill up the space between the larger and smaller pipe, entirely preventing the escape of oil. The three-quarter inch orifice was then further reduced by closing it to one-eighth of an inch, and through this the oil flowed into prepared troughs leading to large tanks.[5] They reduced the flow to 600 barrels a day, but it was estimated that a million barrels were lost, draining into the Sydenham River and finally ending up in Lake St. Clair some 24 miles away.

John D. Noble owned a fleet of ships on the Great Lakes. Standing at dockside in Kingston, watching one of these ships being anchored, he became angry. He was a man who kept his fleet well-painted and spotlessly clean, and this once-white schooner was covered in a black, slimy substance. He hotly demanded that the crew tell him what had happened. The skipper explained that a wild well had blown out of control in Oil Springs, and oil had leaked into the St. Clair and Detroit Rivers. Noble sold his boats and rushed to the oil field where he struck it rich and stayed a lifetime.

It is estimated that ten thousand fortune seekers flooded Enniskillen within a few months. Oil Springs boomed. The main road was planked for a mile and a half with a two-inch thickness of white oak; omnibuses ran every five minutes from one end to the other. Kerosene lamps on ornamental posts lit the streets, the first streets in the world to be fitted with this kind of lighting. Twelve general stores sold everything from cough mixtures highly laced with opium to spittoons and metal pee-pots. Nine hotels supplied crowded rooms, meals, hooch, and burlesque shows. An American tried to buy one frame hotel, standing on half an acre of ground, for $9000, but the owner refused to sell.[6]

Prostitutes trooped across the border from Michigan to get in on some of the action. There's a

The Oxford House was one of the almost dozen hotels erected on the main street of Oil Springs in the 1860s.

An underground storage tank rebuilt on the site of the Oil Museum of Canada.

story that has come down through the years concerning a prostitute nicknamed Christy-Bootjack. Christy lived in a log shack back in the bush, far away from the maddening crowd. At various times in her life, she lived with different men who turned a blind eye to the clientele who found their way to her door. However, one live-in boyfriend did something about this troop of hard oilers with their hard talk and dirty, leather boots. He invented a bootjack that he put at the door, and he insisted that all callers use it to remove their boots. He was simply fed up with climbing into bed between torn, oily sheets.[7]

Hugh Nixon Shaw celebrated his fiftieth birthday two weeks after his wild well blew in. He was older than most of the fortune seekers, but he was healthy and enjoyed his new status as a Very Important Person in Oil Springs. Now that he was an operator like Fairbank, he also grew acutely aware that storing, refining, and marketing his gold brought new headaches.

After controlling the well, he filled every available wooden storage tank he could beg, borrow, or steal, then he decided to dig an underground tank. The underground storage tanks, introduced the year before, were unique to Oil Springs. The blue clay soil that soaked up water like a sponge, making it a nuisance in wet weather, was perfect for oil storage. Shaw dug down 60 feet into the clay, making the hole 50 feet in diameter. He lined the upper 15 feet with curved wooden slats, and made the roof of logs. The average capacity of these tanks was 8000 barrels. As John S. Ewing has observed, "they were impervious to moisture and to evaporation, and they were patently fire proof. . . their lives were long, and some were in use until well into the 1900s."[8]

Shaw built a still and began to sell kerosene and crude oil, but he and Fairbank learned a disturbing lesson from this wild well. As production went sky-high, selling prices fell miserably low. They would wrestle with this price problem for years to come.

RACE RIOT

At the time that Shaw's well blew in, there was a rich payload of gold found at Barkerville, British Columbia, which kicked off the famous Cariboo gold rush. This drew a diverse mix of miners and adventurers into the once fur-bearing district, but hard oilers did not join the stampede. There was too much excitement and promise of wealth in the oil region.

Thousands of squatters descended on Enniskillen, and as J. Harvey Johnson, an eyewitness, has noted, "there were many fights over ownership of lands." A policeman was busy trying to keep order, and a Justice of the Peace, a man named Harrison, was much admired for overlooking a lot of sins if the sinner shared a good joke with him. Johnson recalls that he was famous for saying, "One dollar and costs and time to pay."[1] Harrison made it known in a subtle way that with a little gold in his palm the "time" could go on and on into infinity.

The policeman soon learned that the long arm of the law was not long enough. In March 1863, an outright war erupted with a settlement of Negroes in the eastern end of the village. The *Sarnia Observer* *Advertiser* reported "A Disgraceful Riot at Oil Springs."[2]

From 1840 to 1860, thirty-thousand Black slaves rode the underground railroad from the United States to Canada, and hundreds of families chose to settle in Dresden, 18 miles south of Oil Springs. In a CBC broadcast in 1967, Dr. Martin Luther King, Jr., drew attention to the importance of this "railroad" to freedom:

Deep in our history of struggle for freedom, Canada was the north star. The Negro slave, denied education, de-humanized, imprisoned on cruel plantations, knew that far to the north a land existed where a fugitive slave, if he survived the horrors of the journey, could find freedom. The legendary underground railroad started in the south and ended in Canada. The freedom road links us together. Our spirituals, now so widely admired around the world, were often codes. We sang of 'heaven' that awaited us and the slave masters listened in innocence, not realizing that we were not speaking of the hereafter. Heaven was the word for Canada, and the Negro sang of the hope that his escape on the underground railroad would carry him there. One of our spirituals, 'Follow the drinking Gourd,' in its disguised lyrics, contained directions for escape. The gourd was the big dipper, and the north star to which its handle pointed gave the celestial map that directed the flight to the Canadian border.[3]

However, Canada was not always "heaven," as was demonstrated in 1863 in Oil Springs.

After Shaw's well captured headlines around the world, hundreds of Negro people were drawn to the oil boom. Families from Dresden, Chatham, and points south trekked north, walking, riding, getting there anyway they could. Almost overnight, a fair-sized settlement of these laborers sprang up on the outskirts of Oil Springs, and the colony expanded daily.[4] The Blacks worked more cheaply than other laborers, and it wasn't long before hate and anger began to erupt in the ranks of the established white community.

The climax was reached when a Black shoved Mrs. Justin Bradley, the wife of the proprietor of the Royal George, a well-known east-end boarding house, off the sidewalk. An indignation meeting was called in a little cooper shop near the boarding house, and the unanimous decision of the assembled men was to give the village's newest populace orders to get out.

Early one Saturday evening, a large crowd of men gathered on the main street at the eastern end of the village, formed into line headed by a big six-foot 225 pound Scotchman, and started for the colored people's camp. Marching with military precision, they were joined in front of the Commercial Hotel by the west-end contingent, which swelled the ranks of the marchers to about six or eight hundred men.[5]

In true vigilante style, the mob carried hefty tree branches, iron pipes, lighted torches, and oil to pour over buildings. Colored men had brought their wives and children with them, and had built log houses and frame shanties in which to live. They had stacked hay into bundles for their horses, and these hay stacks were lying around in the open fields near their living quarters. The marchers swooped in, saturated the buildings with oil, set fire to them, then torched the hay in the fields.[6]

In an article headlined "Racial War" published in the *Petrolia Advertiser Topic*, Charles Watten recounts the events of that evening:

The quiet of the night was filled with the screams of frightened children, the wails of the womenfolk, and the shouts and curses of men run amuck. The whole camp became a blazing inferno, through which figures raced madly to escape cremation. . . . The lust for destruction conquered all reason, and the raiders became veritable madmen as they milled about on their errand of devastation.

In a very short time, the village policeman arrived at the scene . . . and special constables were sworn in. Several of the leaders in the riot were arrested, but having no jail accommodation for them, were soon at liberty again. Only two of the hundreds participating in the raid were arrested and jailed — Malcolm McQuinn and John Lavill. They were sent to Sarnia for trial, and were sentenced to Kingston Penitentiary for three years. McQuinn was pardoned in a short time, but Lavill served his full term. . . . Years later when the village had passed in and out of the boom era, the villagers were amused to see a tall, broad-shouldered colored chap marching in from the west with a gun over his shoulder. He was the advance guard of the Michigan Central Railway gang, ready to go to work for the railroad. . . . He was making sure that he would not find himself in the same unfortunate position as his former hapless brothers — hence the gun.[8]

Phil Morningstar, descendant of one of the early pioneers and a family member of the still operating Morningstar oil company, recalls that "there was a hanging when a Nigger, that people called 'a Black Gypsy,' was told to 'get out of town,' and when he didn't, he was hung from a tree. It's something Oil Springs wants t' forget, but it happened. There was a lot of drinking done back then. There was nothing else to do, with mosquitoes eating you alive in summer, and mud everywhere, and cold and snow in winter. . . . Bootlegging went on. Liquor came from Windsor. Corn whiskey. A Belgian bootlegger, a woman and her husband, lived just outside of town. A couple of hard oilers went out to order liquor from her, and she dumped diaper water out of a big wash tub, and didn't clean it before she dumped in newly-made beer and began to measure it out for the men. They left, needless to say, without the brew."[9]

CANADIAN OIL ASSOCIATION

Despite this social turmoil, Hugh Nixon Shaw began to refine his crude. He invented and patented a still of his own that consisted of two sugar kettles placed one above the other, and from the top, a pipe was connected to a condenser. The vapor passed up this pipe through a fine iron screen, and then a brass wire mesh which Shaw claimed sifted out impurities. From this iron pipe connecting the still to the condenser another small iron pipe passed through the roof to the open air and allowed volatile vapors to escape. The remaining vapor became liquified in the condenser and was drained to a collection tank. Shaw claimed to make 50 percent of illuminating oil by this procedure; all else was lost. "The illuminating oil was transferred to a tank where it was agitated in a sulphuric acid bath," Edward Phelps explains. "It was then neutralized by washing in caustic soda to remove the evil-smelling sulphur."[1]

This method, however, did not remove much of the sulphur. These early refineries were built cheaply, and they were imperfect. They were also risky business, constantly exploding or catching on fire. John Henry Fairbank tried Shaw's refining method with little satisfaction, as he noted in his diary: "About as miserable a day as I ever put in, run till dark and quite fully realized that I won't run a damned leaky old kettle that acts as if it would 'go up' at any minute... don't want to become as nervous as an old maid, and feel like a coward all the time. I'm down on the thing, and won't stand it anyway. Can stand work as well as anyone, but damn a leaky still; them's my sentiments. A bright day; oil settling."[2] He made another note in his diary a month later: "Fixed pump. Tank nearly full. Shaw peddling oil; he is a big ass, would do a smashing business at selling molasses candy and peanuts."[3] According to knowledgeable oil men today, the Shaw whom Fairbank refers to here was not Hugh Nixon Shaw, but rather another man named John Shaw, not related to Hugh Nixon.

Oil had become big business, for kerosene was in demand all over North America and in England. British money, lavishly mixed with American currency and some Canadian silver, was used in everyday business transactions. For loans and banking, the oil men usually went to Sarnia where the Bank of Upper Canada was located until 1866, then was replaced by the Bank of Montreal.[4] Frontier farmers, often in debt, mistrusted the banks and businessmen of the towns. The Bank of Upper Canada was founded in 1821, but farmers thought that this banking business was a ploy by public officials to help the government grow rich on the monies that people were crazy enough to invest. Many oil men identified with the farmers, and some of them hid their money in mattresses, old shoes, and even old wells, as hard oiler Bruce A. Macdonald recalls.[5]

There was one private bank in Oil Springs in the early 1860s. Leonard Vaughn, an American, cranked up this bank, but it did not last long as small operators trusted private banks even less than they trusted government-sponsored ones. Vaughn had farmed near Titusville where he watched the "Colonel" set up his rig to bring in the second commercial oil well in North America. In his early twenties, Vaughn was fascinated with the oil business, and when the American Civil War drew many men away from Titusville, he decided he would go to the Canadian field. In 1861, he was drilling on land adjacent to Fairbank's land, and the two men became friends. At a later date they became partners in a private banking business that would prove popular and profitable.

Shortly after Shaw's well made its spectacular debut, a steam-powered rig drilled the Bradley gusher, and a month later Black and Mathieson of Sarnia brought in the greatest flowing well in Ontario's history, a gusher that sent oil spiraling skyward at 7500 barrels a day. In his history of the Union Gas Company in Southwestern Ontario, Victor Lauriston notes that this wild well "defied all efforts made to control it, and flooded the lowlands to a depth of a foot or more so that men went about with jumping poles, skipping from log to log."[6]

Before October arrived, thirty-five wild wells had burst upon the scene, and the selling price of a barrel of crude had fallen from one dollar to ten cents. As Dr. Alexander Winchell at the time reported, "no less than five million barrels of oil floated away on the waters of Black Creek."[7]

Each day, five hundred teams left Oil springs for the railroad terminal in Wyoming. Land prices sky-rocketed. One Chicago company paid $14,000 in gold for eight-and-a-half acres, and offered to buy the adjacent eight-and-a-half acres for $20,000, but the owner refused to sell. Another American company paid $80,000 for Lot 21 Concession 1. Twenty-seven refineries churned out kerosene.[8]

The English market had opened up the year before, but it cost $8.83 to send one gallon from Oil Springs to Liverpool via New York. A barrel usually held 40 gallons, and shipments were made from 1000 to 35,000 barrels.[9]

Fairbank decided to do something about the low price of oil. He contacted the operators and asked them to band together to form a company to set the price at one dollar a barrel, crude. It was not an easy feat to get the producers to join together for a common cause. They liked to do business on their own and set their own prices. Also, they did not trust one another. Fairbank once said that "if men were not so selfish — the short-sighted kind — and such fools, they could sell just as much at $1 as at any lower price, but the arrangement must be unanimous and such a thing is difficult."[10]

He succeeded in corralling the operators under the banner of the Canadian Oil Association, governed by four officers and nine directors, with

The Oil Well Supply Store was one of many business enterprises to spring up in response to the oil boom.

Fairbank in charge of all. It was organized as a one-year partnership among oil producers. It was the first effort to bring order to the fledgling industry, but the association was not without its problems. Arthur B. Johnston, in his book recording the memories of his father, J. Harvey Johnston, points up these problems:

There were two wells on a lot belonging to William Woodward on the side hill. . . . One of these wells was producing good at the time that the oil producers organized to curtail production and store their oil to obtain a fair price. This man, however, kept on producing and selling. One night some men went to the well; they broke the top of the pump off and filled the hole with old iron and anything else they could find. He hired my father to clean the well out, and after a great amount of work they got it cleaned out down to the big lime, a few feet from the vein. That night my father received a letter which was signed by many oil men, warning him to quit work at once under the threat of no more work or contracts from the oil producers. The man then got an outsider to drill a new well beside the [sabotaged] one, but this proved to be a dry hole.[11]

While pioneers struggled to bring some form of control to the petroleum industry, Canadian minerals were receiving acclaim an ocean away.

ROYAL HONORS

The Crystal Palace in London, England was an epochal building of domed glass ceilings and wrought-iron fixtures, and it was aglow with gas lights, color, and orchestral music for the 1862 world exhibition, the second since the Paris International in 1855 where William Edmond Logan and Charles Nelson Tripp were celebrated. But the pomp and splendor of the Paris fair seven years earlier were not there. Although Prince Louis Napoleon and his beautiful queen Eugenie visited the exhibition on one occasion, they did not reign over the festivities as they had in Paris, and the dazzle and excitement of the French court were sadly missing. Also, a few months before the fair opened, Queen Victoria's husband, the handsome, dashing Prince Albert, died of typhoid fever. The queen went into deep mourning, and this put a pall over the festivities.

The Civil War, too, was in full swing in the United States, and American exhibits were noticeably absent. The fair came off a poor second to the war in newspaper coverage. Logan (now Sir William Logan after receiving a knighthood six years before) was a quiet, subdued visitor compared to the high profile picture he created at the Paris Exhibition.

In spite of all the strikes against it, the fair drew six million people to see 29,000 exhibits. The Toronto *Globe* reported Logan and his associates in the Canadian Geology Department received a gold medal "for an admirably prepared selection of specimens illustrating the mineral resources of the province." J. Sterry Hunt, the geologist who in 1849 first reported oil-findings in Enniskillen, received a gold medal for "the instructively described series of the crystalline works of Canada, and his various published contributions to geological chemistry." [1]

James Miller Williams was, however, the most honored Canadian. Anticipating these awards, a Toronto reporter conjectured: "One of the medals for mineral products will go, I believe, to Mr. Williams who first bored for oil."[2] In fact, Williams was presented with two gold medals by the Duke of Cambridge, representing Queen Victoria. The first one was given for "introducing an important industry, by sinking the artesian wells in the Devonshire Strata for petroleum"; the second one was for "an extensive exhibition of the derivatives of petroleum."[3]

Back in Oil Springs, the petroleum industry was booming. The selling price of crude rose to 75¢ a barrel, which was, as the *Globe* reported, "an advance of 25¢ upon the [Canadian Oil]Association's price that was first established."[4] Before the end of the summer, seventeen vessels loaded with 35,000 barrels of Canadian oil, were on their way to Europe. And the domestic market began to grow rapidly.[5]

Hugh Nixon Shaw's oil at last found its way to England and Europe, thus opening up wider markets for him. He had invented a method of extracting paints and dyes from oil that further expanded markets for by-products.[6] His glory days were to be short-lived, though. In January 1863, salt water began to choke out the oil in his great flowing well. On the 11th of February, he asked two helpers to lower him into the well by operating a windlass. A pipe had broken 15 feet below the ground, and he wanted to retrieve this pipe. Early oil men were often lowered into wells to perform maintenance. After Shaw had been lowered and had picked up the pipe, he called to the men to haul him up. His helpers heard him drawing several long, labored breaths, and before they could get him to the top, he fell back down into the oil and disappeared.

The attending physician, Dr. S. Street Macklem, on viewing the body, "found near the base of the occipital bone, a short transverse wound penetrating to the skull, but not fracturing." This, however, was not a mortal wound. At an inquest later, the jury rendered a verdict that Shaw "came to his death from asphyxiation caused by poisonous gases inhaled whilst in the well."[7]

The oil field was in shock. Fairbank, speaking no doubt for himself, Williams, Vaughn, and other hard oilers, noted in his diary: "Poor Mr. H. N. Shaw, drowned in his well today, in him I have lost one of my best friends in Enniskillen. A good man and most obliging neighbor; sad, sad, sad calamity."[8] An obituary in the *Christian Guardian* eulogized this devout hard oiler:

Mr. Shaw has long been in the enjoyment of religion, and in a diary kept when a young man, he gives a very interesting account of his conversion, which took place in his 21st year. He then united with the Wesleyan Methodist church, and speaks in the highest terms in praise of the various means of grace, especially the class-meeting, and for 32 years he continued to love and attend these services, and was found a consistent and devoted member of the church at his death. It would appear as if the Lord had been for some time past more fully preparing him for his sudden departure, as one of the last books he had been carefully studying was "Foster on Holiness," some pages of which he had as carefully marked. He was kind and much beloved by those who really knew him. [9]

Within two years, his wife Anne died of a broken heart. Her obituary in the *Christian Guardian* noted that "the death of her husband was fatal to her; from that date to the date of her own death, her life was the willow drooping and weeping." [10] Within two months, Shaw's oldest son, Bartholomew, aged twenty-nine years, who had operated the store at Cooksville, also died. He was buried beside his mother and father in the little country cemetery near Davenport United Church, an edifice that his grandfather, Bartholomew Bull, had helped to build. There are no records to show what illness or accident claimed the life of young Shaw, but a reporter said that "he was a beloved son and tender to his sisters. His father's death and mother's decline brought him much responsibility." [11]

Hugh Nixon Shaw could not have made a fortune from his wild well. In fact, after all his debts were paid, there was probably just enough money left to bury him.

OLD FAIRBANK

Within a month of Shaw's death, salt water choked off the oil in twelve flowing wells. With this dearth of crude, the selling price of oil rose to $3 a barrel.[1] Demand outran supply as the surplus of oil vanished. This turn of events accomplished what the Canadian Oil Association was unable to do. The association disbanded. As Edward Phelps comments, "Whatever its influence, it was an early triumph of organization within the fragmented and competitive oil industry and served as a model for future combinations. . . . It followed only a few months after a similar combination formed in the Pennsylvania oil region for the same purpose."[2]

John Henry Fairbank continued to "milk" his "Old Fairbank" well, which he referred to fondly as "my old pet" and "my first love."[3] But, his finances were in a bad state. He drilled another well and hit a dry hole, borrowed money from his father-in-law to cover debts, and was sued by Williams for $281.30, the monies still owing on his oil land. He lost the case in Lambton County Court, and had to pay the debt forthwith, along with court costs of $52.81.[4]

He built a cabin 12 by 16 feet, but Edna, wrestling with a farm that had produced poor harvests the year before, called it "a house without hope."[5] He spoke many times in his letters about being lonely, and he wanted her to join him in Oil Springs, but she refused. In one letter, she said that "if you should be taken from us, me in no business, and we so in debt, true there would be sufficient to pay debts and a home left, but we must have something to live on, and my business [farming] once given up would be hard to take up again. I have not the heart nor the strength to take up the subject at every point as you have. God knows, I would do anything to please you but give up my work, and I think that nothing would give more love for my husband and children than to continue on in the road I have been treading."[6]

In a later letter, speaking of visiting with a friend and talking of oil, Fairbank apologized: "Well Ed, you see, you ain't here and the oil is all the time, and my relations with it are very intimate."[7]

Fairbank's mother, Mary Oliver Fairbank, eventually arrived, with four-year old Charles, to live in the new log house in Oil Springs. On a Sabbath evening, he presented to Edna a picture of domestic bliss:

I came in late from Petrolia last evening, did not rise early in the morning, but when I did and had cleaned up, away downtown went Charley and I, and got Mama's letter and pictures and papers, not forgetting the almanac, then accompanied by Mr.

Kerby we returned to our shanty and dined from a splendid piece of roast beef, potatoes and onions which leaves me with a strange feeling of fullness. Indeed, I feel just like stretching out and doing nothing. Mr. Kerby has gone to his quarters, Charley has had a long frolick, mother sits knitting."[8]

Charley wore a tag pinned to his clothing at all times, giving his name and the exact location of his home, a protection in case the child became lost. But Charley did not get lost, and his grandmother, then in her seventies, adjusted well to pioneer living.

Other gushers replaced the ones that gave up the ghost, but the days of the wild, flowing wells were nearly over. This development did not, however, spell the end to the boom. Men continued to drill in the hope of hitting a big one, and shallow wells that yielded 30 to 100 barrels a day continued to produce.

In June 1864, the selling price of crude jumped to $5 a barrel, and Fairbank managed to snare the biggest contract of his career, 1000 barrels ordered by a refining company in Woodstock. Shortly after this he sold "Old Fairbank" to Abraham Farewell of Oshawa who was in partnership with four American capitalists. Upon instructions, he went to Sarnia where a bank, on behalf of the buyers, paid him $6000 in gold for the well.[9] With the gold sacks weighing heavily in his arms, he "danced" all the way back to Oil Springs. With this stake, Fairbank refined the methods of drilling for oil and the methods of refining oil for market.

DRILLING AND REFINING TECHNOLOGY

Early refineries were big, cast-iron receptacles that looked like ink bottles or kettles. These "distilleries" or "stills" were placed on brick mounts that contained fire boxes fueled by coal or wood. There were no controls on these early stills, and, as the oil changed to gases, and the gases were condensed to liquids and were separated into gasoline, naphtha, and kerosene, the early refiner used his own judgment (a by-guess and by-God theory) to shut off one from the other. In this way, no two kerosenes were the same, and a few, too heavily laced with gasoline, exploded in lamps.

Many by-products of refining were a nuisance, especially gasoline. Nobody wanted to store gasoline because of its volatile nature, and yet there were no known uses for it, so oil men had to get rid of it. They resorted to the easiest method by dumping it in the ground or pouring it into nearby creeks. This method of elimination proved hazardous.

In 1859, James Miller Williams built a still on the bank of Black Creek. One day he dumped half a dozen barrels of gasoline into the stream, and a few hours later a storm erupted pouring tons of water over the land. The creek rose, and gasoline, floating on the water, reached the fires under his still. The explosion ripped stalwart trees out by the roots, and the creek became a raging inferno for a quarter of a mile. A few years later, lightning struck gasoline as it floated blithely along the creek, and the avalanche of rain that followed did nothing to contain the holocaust. People ran in all directions, taking with them their most precious possessions, but the fire remained entrenched in the creek, catching only the stills that were built along its banks.

Sometimes, fire erupted within a still when flames in the brick furnace ignited gasoline vapors, causing violent explosions. Refinery operators were always poised, ready to run if the apparatus started to agitate or leak in a strange way.

By 1864, hard oilers began to specialize, as G.A. Purdy in his history of the petroleum industry notes: "Some men became drillers; others produced crude oil; and still others transported it. The refiners also specialized. Most concentrated on making kerosene, but others produced only lubricating oils, waxes or naphthas. Oil products were sold through existing jobbers who normally handled hardware, foodstuffs and the like."[1]

Refiners were well aware of the hazards of running a still, and in the early days, they, like the producers, wanted to regulate refining methods and stabilize the selling price of a barrel of oil. In 1861, the Petrolia Refining Company was formed by a group of refiners with these aims in mind, but the association collapsed within the year. The refiners were difficult to organize. They were jealous of one another, secretive about their activities and extremely competitive. As Edward Phelps comments, they "should have been easier to organize than the producers for their ranks numbered only a tenth of the latter. Yet, the refiners never achieved the same degree of solidarity as the producers." At a later date they formed The Refiners' Association "to regulate the price at which illuminating oils are placed on the market, to control prices and to promote trade."[2] This association collapsed, too, within a year, but before this they petitioned the government to appoint an inspector "to examine the quality of refined oils and reject those which were not of good quality."[3] It was not until 1868 that the government fixed the "flash point" of illuminating oils at 115 degrees Fahrenheit and a standard specific gravity test was set for the refining process.

Max W. Ball defines "flash point" as follows: "When oil is heated enough . . . vapor begins to form, collecting just above the surface of the liquid oil. If a flame is brought close to the surface of the oil as it is heated, at a certain temperature enough vapor will form to ignite from the flame, causing a momentary explosion or 'flash'. The temperature at which this takes place is known as the 'flash point'."[4] Thus, with a standard flash point test, it was assured that volatile substances like gasoline would not get into the kerosene.

Fairbank allied himself with the producers, but, like Williams and other hard oilers, he also operated a refinery and he could identify with the problems of refiners. He re-invested monies from the sale of "Old Fairbank" into the purchase of lands, and he drilled more wells that were slow but sure producers

A primitive refinery or still located on the grounds of the Oil Museum of Canada.

like his "First Love." By 1865, he was a very important person in Oil Springs. People consulted with him on land speculations and all aspects of the oil business. Unlike Charles Nelson Tripp, he was keenly interested in politics. In 1861, the year he arrived in Oil Springs, Alexander Mackenzie represented the county of Lambton in the Legislative Assembly. Mackenzie was a familiar figure in Enniskillen when campaigning for votes, traveling by horse and buggy, speaking in school houses and hotel saloons. As Liberals, he and Fairbank were on the same side of the political fence; early in his career, Fairbank became friends with the rugged, Scottish mason who would later become the first Liberal prime minister of Canada.

Getting their oil out to market had always been a problem for the pioneers in Enniskillen. Mackenzie was the push behind getting a group of Sarnia citizens together to build a plank road from Oil Springs to the port town. A few years earlier, in 1862, a private company had built a plank road to Wyoming, thus replacing the old "hell hole." But the 20 odd miles of plank road from Oil Springs to Sarnia in 1864 was like a miracle. The railroad charged three times as much to carry a car of oil as a car of dry goods, and, once the road was built, oil men could put their barrels directly on boats bound for England and Europe, thus cutting out railroad freight.

Plank roads were a Canadian invention of the mid-1800s, and their reputation as the ultimate in comfort and economy spread quickly to the United States where the idea was eagerly adopted. Wooden planks made of four-inch thick oak were placed across the roadway and secured with spikes to stringers. One mile of plank road cost 1000 pounds to build. There was a lot of wear and tear on these roads by horses' hooves, wagon wheels, and the changing elements.[5] The roads were only 16 feet wide; although turn-a-bouts were supplied here and there to facilitate two-way traffic, they were not always at the right place at the right time. Many wagons had to get off into the mire at the road-side to let somebody else go through, and this caused problems in getting back on to the planks later.

Within two years of construction, both plank roads were in need of repair, and the companies that built them were broke. They were turned into tollways, but revenues were not sufficient for repair work, and eventually the plank roads became impassible. And as John Maclean in an article published in *Imperial Oil Review* observes, "the teamsters, a hard-muscled, hard-drinking, devil-may-care gang took the better part of a week to make the round trip to Sarnia, and complained that they spent more than they earned in the four taverns that lay on their route."[6]

As early as 1862, when melting snow swelled Black Creek, one ambitious oil company floated 1000 barrels of crude the 24 miles from Oil Springs to the St. Clair River. Only one barrel was damaged beyond repair; the cargo was loaded on a waiting steamship bound for England. Yet this was not an easy way to get oil out to market. Fairbank tried the same experiment a year earlier. Four thousand barrels were rolled into the creek one morning in the fall of the year, and a gang of men pushed them along, mile after mile. One hard oiler recounted the journey:

The creek was filled with saw logs and timber and the men were often up to their armpits in the water. On the second morning . . . the creek was sheeted with an inch of ice. "May as well give it up; can't do it," Doc Aikens exclaimed. Fairbank came up brandishing his handpike and saying, "Dash you, Doc, if I hear can't do it from you again, I will down you with this." The men worked like heroes in the stream and out of it, and the cargo got down to deep water and was put on board a vessel. But the feat was in vain, for the ship was lost in crossing the Atlantic.[7]

The men struggled to control the price of a barrel of oil, refining methods, transportation problems, and the high cost of pumping. The great flowing wells did not need to be pumped. Nature supplied the power for this, but the shallow wells, similar to water wells, needed individual pumps, either hand pumps or steam pumps. As Wanda Pratt and Phil Morningstar explain, the use of "steam engines on each well . . . was prohibitively expensive for most oil men. Finally, Fairbank figured out a way, almost by accident, to hook more than one well to an engine for power. He had two wells near each other that needed to be pumped. He had a steam engine on one, but didn't have room to install another engine. He took an extra

An old tank wagon abandoned in a field at Oil Springs.

A walking beam pumping mechanism at work in a jerker line system at Oil Springs in 1993.

pole and hooked it on to the orphan well, and that was the beginning of the jerker line system."[8]

By this method, one five-horse power steam engine could pump twenty-five wells. Fairbank used a field wheel, powered by an engine in a nearby pump house. The wheel rotated back and forth driving jerker lines to all wells. As Pratt and Morningstar explain, "these jerker lines were 30-foot long wooden poles, two inches by two inches, hooked together lengthwise and running from the field wheel to all the wells needing power. The jerker lines were hung from wood poles set in the ground along their path. Wire hangars were hooked to the top of the pole and on to the jerker rod, both holding it up and allowing it to swing back and forth. Careful planning was needed to insure that there was equal weight on opposite sides of the field wheel."[9]

Although Fairbank introduced the jerker-line system of pumping to Canada, it was well known in Europe in the 16th century. Called *Stangenkunst* (rod work with crank), it was used in mining in German-settled areas, to transmit the power from water wheels to pump mines at a distant location.[10] Still, the jerker-line system was unique to Oil Springs, and today it is still being used on several hundred of the old historical wells that have kept pumping since the early boom years. The jerker lines squeaked and hooted softly as they moved back and forth, two lines, side by side a few feet off the ground. Hard oilers became familiar with this gentle hooting, and when or if it stopped, they heard it like a cannon exploding. Silence meant trouble.

Well drilling technology also advanced during these years, as Pratt and Morningstar explain: "By 1865, steam engines were used for drilling, and they were moved from one well to another behind the new style Canadian pole drilling rigs. These rigs had four poles, stabilized by cross beams."[11] An auger replaced the old spring-pole method of drilling; this auger turned, making a hole in the ground, and it was powered by steam or by a horse plodding in a circle. The steam-engine rigs were 30 to 50 feet high and equipped with runners, much like sleds, and they moved around muddy fields and roads with ease.

Hard oilers welcomed newcomers who were looking for work by sending them to the top of the beam of one of these rigs as it was being moved. When the newly-arrived "dude" finally managed to climb to the top, the driver cracked his whip over the horses' backs, propelling them suddenly forward. If the new man fell off the rig into the mud, he was told he was not good enough to work in the oil patch. If he held his place on top, he was hired forthwith.[12]

Also in 1865 the American Civil War ended. Oil Springs felt the impact of this armistice when there was a rush of ex-American soldiers to Enniskillen. Among them came men who would have a powerful impact on the Canadian refining business, much like John D. Rockefeller in the U.S.A.

Field wheel, jerker lines, and pump house reconstructed on the grounds of the Oil Museum of Canada.

JAKE ENGLEHART AND WILLIAM H. MCGARVEY

In 1865, John D. Rockefeller put his foot on the first rung of the ladder of corporate success when he became sole owner of the largest refinery in Cleveland, Ohio. Within a year he built a second refinery, and, in need of markets, he organized a small company in New York to take care of the export of his kerosene. In that year he achieved phenomenal success; his sales for oil amounted to two million dollars.

Rockefeller was born in Richford, New York in 1839, and when he was a small child, his father William moved to Cleveland where John was educated at public schools. By age sixteen, he was working as a bookkeeper in a produce firm, and in 1862 he became a partner in a small refining operation. In the next three years, he bought out partners until he was sole owner of the big still. This was the beginning of a campaign that brought order and control to the chaotic refining industry in the U.S.A. at the expense and seething anger of small refinery operators. It was the beginning of Standard Oil Company.

Canadian refiners were well aware of Rockefeller, and, like small operators of stills south of the border, they did not like his underhanded business tactics. At a later date he guaranteed large shipments of oil to the railroads, and in return the railroads offered him rebates. This put the small refiner at a great disadvantage; he was already struggling to eke out a living, and a big octopus like Standard

was cutting deals with the railroads that would make it even more difficult for him, the small guy, to stay alive. The general public cried out their wrath against Rockefeller, and small refinery operators devised ways to retaliate.

A man who became a tycoon in Canadian refining operations also hailed from Cleveland. Jacob Lewis Englehart, who was called "Jake" by many friends, was the son of John Joel and Hannah E. Englehart. Born in 1847 in Cleveland, Englehart was educated at public and high schools there, and by age fourteen was working in his father's store, learning the intricacies of the mercantile business.

Rockefeller was a strong Baptist and the Engleharts were Jewish, but Cleveland was a small city in those days, and it is possible that young Jake knew John D. He was certainly well aware of the oil distilleries on the outskirts of the town, but his first job was selling booze for a whiskey distiller in New York.

After the Civil War, when there was a rush of Americans to Enniskillen, Englehart joined the stampede. He set up a firm, Englehart & Company, in London, and combed through the oil fields in search of small producers who would be willing to sell their crude at a cheap rate. Unlike the little man who started to refine on a shoe-string, Englehart always seemed to have enough capital to cover his personal and business costs without worry. Hard oilers wondered where he got the money to set himself up in business, but Englehart did not divulge any information about his New York connections. He avoided press interviews, and, during his lifetime, he managed to keep his early beginnings a secret. It is doubtful that when he married twenty-one years later, his wife was aware that he had used laundered money to get started in the refining business.

As Hugh M. Grant in an unpublished study of the petroleum industry notes, Englehart was a salesman for Sonneborn, Dryfoos and Co. in 1865-66, "a New York firm of 'Whiskey Rectifiers' composed of Solomon S. Sonneborn, Abraham M. Dryfoos and Leopold Beringer.... Sonneborn had spent five years in Europe as a director of the American Rubber Company in which he had amassed a substantial amount of capital. Dryfoos had migrated from Germany and opened a trade in linen goods in Philadelphia with his two brothers. When their Philadelphia store failed, they undertook a 'cotton speculation' in Metamoras, Mexico, leaving behind in Philadelphia several debts and a reputation of unreliability. Leopold Beringer had been engaged in a tobacco business with his three brothers, but had also made a substantial amount in the illegal whiskey trade and was similarly deemed to be 'a shrewd, unreliable man.'"[1]

The R. G. Dun Credit Agency in New York noted that Sonneborn, Dryfoos and Co. "has the reputation of handling more illicit goods in a given time than any other house in the trade here."[2] The partners were charged with violating U. S. revenue laws, "but

Jacob "Jake" Englehart was a founder of the petroleum refining industry.

by management and by some of them leaving the country, they were enabled to escape the penalties of the law." However, many old indictments of this kind were brought to light, and the government was determined to pursue them.[3] At a later date, a Dun reporter wrote: "They established themselves in Canada in 1869 and the move was regarded as the establishment of an Asylum for the men who had hitherto been employed illicitly here, and for the investment of means which might otherwise [have] been pursued by the U.S. Government."[4]

Englehart is reported to have been in Canada at the age of nineteen, which would put him in operation in London in 1866. Grant states that "the importance of this New York connection to the Canadian petroleum industry cannot be overstated."[5] In fact, this connection eventually turned refining into a powerful corporate business.

In the beginning, though, Englehart's biggest problem was that he looked like an overgrown school kid, no beard, not even a down-like fuzz on his face, and therefore too young to be trusted. Also, word spread quickly that he was "a city slicker from the States."[6] Nobody had heard of him, and competitors warned small producers to be wary of this new, young upstart.

While trying to drum up business, Englehart lived in London at the Tecumseh House Hotel. One evening as he was going out for a walk, he noticed two wooden woodcocks on the hotel owner's desk. They were decorations, spigots taken from a whiskey barrel. When Englehart started to walk down the road, he met the head of the game protection association. Ian Sclanders in his biographical sketch of Englehart recounts what then happened:

On an impulse, Englehart informed the game protection man that he suspected the hotel owner at the Tecumseh was breaking game laws as he had seen two woodcocks in his office. Without checking up, the president of the game association rushed off and laid a charge against the hotel keeper. Englehart was summoned as a witness, and when called to the stand, he confirmed his statement that he had seen two birds — woodcocks — on the proprietor's desk. "I have them here," he said, producing the wooden woodcocks from his pocket. Everybody roared with mirth — except the president of the game association. And, in the back concessions of Enniskillen, where a good joke was appreciated and hunting restrictions were disliked, the incident cemented Englehart's prestige, and his reputation of being a "real man even if he looks like a schoolboy."[7]

His London refinery was built and ready for operation by the fall of 1866, and he had a good supply of crude oil ready to feed it. Although distilling whiskey and distilling crude oil were similar operations, there was also a vast difference, and Englehart was not familiar with the volatile substances cooking in his still. There were so many explosions in one year that the New York company no doubt wished they had never heard of Canada and its oil. But Englehart, like Tripp, believed in his product. He was also stubborn, and he persevered through rough times, pacifying as best he could his New York connections. At a later date, he became, in Canada, the counterpart of Rockefeller in the U. S. A. Although his methods were not always beyond reproach, he fought hard to control fluctuating oil prices, to inaugurate safety measures in refineries, and to amalgamate many small operators into a big company.

Another Oil Springs man was to become considered as the Rockefeller of Europe. William H. McGarvey was a lanky sixteen-year-old when he came to Wyoming with his parents in 1860. Bill's father, Edward, and his mother, Sarah, had emigrated to Canada from Ireland in the early 1840s. They settled in Huntingdon, Quebec, where Bill was born, and in 1860, when they heard about the oil boom in Oil Springs, they moved to Wyoming to open a general store. Edward advertised this business as "the largest general store in Canada," and he carried oil well supplies as well as groceries, wines, and a variety of medicines.

Young McGarvey worked in the store with his father, but as early as 1862 he was tramping through the oil fields learning all he could about the drilling, refining, and marketing of oil. He became well acquainted with Williams, Fairbank, Vaughn, and Shaw; and, although he drilled successfully at Oil Springs, he was one of a new generation of young

men who saw great potential in Petrolia, nine miles north of Oil Springs and closer to the railhead at Wyoming.

The name "Petrolia" was coined by an early postmaster from the words "petrol" and "lea" because of early finds of petroleum and gum beds in the area. In 1860, people called it "Petrolea" but eventually the "e" was changed to an "i". The first wells were drilled in the Bear Creek flats of east Petrolia in 1860 by a man named Kelly. "Although the oil was plentiful, the great quantity of water in the wells made handling difficult," Jean Turnbull Elford explains.[8] By 1861, there were seven oil wells in the area that were producing moderately well, and the Petrolia Refining Company erected a big still. The company was composed principally of men from Boston with a capital stock of $20,000.[9]

In spite of all of this activity, Petrolia took a back seat to Oil Springs for another five years, and in that time McGarvey had a foot in both camps. He owned an oil well in Oil Springs and, in 1866, he became the first reeve of Petrolia when it was incorporated into a village.

Young McGarvey liked stylish clothes and stylish buggies, and when he became reeve he bought himself a leather-cushioned carriage and a fine mare to pull it around the oil fields. He wore suits that were especially tailored for him by a local tailor. Although most of the time, when working around oil wells, he wore old work clothes and knee-high leather work boots, there were other times when he

*William H. McGarvey became
one of the most famous
"hard oilers" in the world.*

got dressed up to go driving in his new buggy. One day he dressed carefully in a navy suit, high starched collar, white vest and highly-polished shoes. He drove past a well site where later the famous Deluge well that spouted 600 barrels a day was drilled. He noticed that John Scott, the engineer, and a helper were struggling to knock the spigot from an oil tank. McGarvey, like Englehart, was always eager to help a friend, and he drove his buggy over to the work site. Scott and his helper noticed the dandy new buggy and the dandy new suit, and they decided to play a practical joke on the new reeve. They would let the spigot out carefully from the tank, and lightly spray McGarvey with black crude. But the spigot had a mind of its own. It flew out, suddenly, sending a deluge of oil all over McGarvey and his fine, new buggy.

The reeve's Irish temper raged for a moment, but he calmed down when he thought the whole episode was an accident. The men began to laugh, and realizing that he looked like a dirty rugged vagrant, he laughed, too. However, the smile faded from his face when he realized that the brand-new leather cushions of his carriage were black and greasy as a result of this "accident."

McGarvey and Englehart became good friends, and each, in his own way, would take diverse paths — often riddled with tragedy —to fame and fortune.

THE RETURN OF CHARLES NELSON TRIPP

In the summer of 1866, Charles Nelson Tripp returned to Enniskillen after an absence of ten years. He must have been shocked at the sight of the metropolis that had evolved with hotels, stores, houses, derricks, mud, and people spread across a treeless plain. When he left, James Miller Williams and a small crew of helpers were working quietly in the bush land. Robins chirped in the trees; geese honked in the marshes. Now, nine hotels did a roaring business and ornamental street lights illuminated the night. Visitors referred to the dazzle of lights at Oil Springs as one of the "seven wonders of the modern world."[1]

An American entrepreneur was building a new hotel, three stories high with two large wings and a long, wide veranda stretching across the front. At the time that Tripp stepped back into focus, painters were decorating the 108 rooms in a bone-white motif, and the outside front in beige. Appropriately called "The International Hotel," the building was equipped with a large, modern dining room, kitchen, and network of hallways with fancy lighting.[2]

Tripp meandered around the oil field, talking to people, and he realized that crews of strangers were reaping big harvests from his old lands. Before crossing into Canada, he had visited Schenectady, and had learned that Henry had kept taxes paid up on three separate lots in his brother's name. On 14 August 1866, he sold one of these 100 acre lots for $7000. On the deed, Tripp is listed as a "gentleman, formerly of the city of Hamilton in the Province of Canada, but now abiding at the city of Ottawa in the said Province."[3]

He certainly must have been happy about this sale. It was a good price for a piece of land at that time. The great flowing wells were now only memories, and drillers lived on hope and speculation that they would strike more gushers. The boom days of Oil Springs were beginning to wind down.

It is believed that Tripp had been in touch with the land agent in Ottawa prior to this sale, and afterwards he hurried back to Ottawa to talk again with the land agent. A reporter, traveling on the same train, met him when he was on his way back to Oil Springs. This writer captured a graphic picture of Tripp and the life he had led since he ran away from the Sheriff of Lambton. The reporter called him "Pratt" throughout the article, but the following week there was a correction, stating that the person referred to in the story was in fact "Tripp."

Headlined under the title "The Original Oil Man of Canada," the article was dated 21 September 1866:

About ten years ago, a man named Tripp, who possessed in a remarkable degree the roving disposition, happened to pass the Township of Enniskillen, and seeing the Gum Beds as they were termed, shrewdly guessed that there was something good to be found there. He accordingly bought 700 acres of land, in what is now the center of the oil district, and after having the gum analyzed at Hamilton, left the country. He went to the silver mines of Mexico and amassed a considerable fortune, from thence he struck over into Texas where, no doubt, he found adventure enough to satisfy even his own spirit and on the breaking out of the late war, took up arms on the side of the South, and fought the Yankees for four years. In the meantime nothing was heard from him, and on the discovery of oil in Enniskillen, his estate was administered and reverted back to the Crown.

About three weeks ago, the rover stepped into a large oil establishment, announced himself as the original old Tripp, and politely requested the *Soidisant* proprietors to clear out. Utterly confounded, these gentlemen handed $10,000 to quit the title and Mr. Tripp set out for Ottawa, where he asked the Commissioner of Crown Lands by what authority his lands were taken away from him and given to others. The answer was in so far satisfactory that Tripp departed in high glee, and when the writer met him on the cars, was on his way to Oil Springs to give a number of wealthy trespassers, as he considers them, "particular sizzors." He expects to realize $200,000 in gold from the transaction, after which he will steer his bark straight to the silver mines of Mexico, introduce valuable machinery, and make his fortune.

Tripp is a remarkable looking man, who might be readily taken for a shrewd, sunken-eyed, hard-faced and eccentric western farmer, and his career is remarkable enough to furnish material for a "thrilling" sensational novel. As far as we could learn he is alone in the world, "Caring for nobody, and nobody caring for him."[4]

Tripp did not tell the reporter that once-upon-a-time he had a wife who seemed to care for him, a wife he deserted. Also, he did not tell him why, originally, he "left the country."

On this second visit to Oil Springs, Tripp stayed only a few days. There is no evidence that he received $200,000 from any further land transactions, or that he gave anybody "particular sizzors." No doubt he visited Williams and they talked of the old, quiet days when they searched for the elusive oil vein and finally tapped into it. After this, he went back to the southern States where he booked into the St. James Hotel in New Orleans. Three weeks later, alone in his hotel room, he died of "congestion of the brain." He was forty-three. There are people in Oil Springs today who believe that Charles Nelson Tripp was a "drunk", hopelessly hooked on the moonshine of the day.[5] There are no records to prove this, but it could be true. His life was certainly erratic.

His brother Henry remained loyal to the end. He claimed his brother's body and buried him in the family plot in a quiet country church yard in Schenectady. An obituary on 2 November 1866 read as follows:

Death of C. N. Tripp. "The Original Oil Man of Canada," as he has been sometimes called, died at

Be it Remembered, That on this day to wit: the *Thirtieth of September*
in the year of our Lord one thousand eight hundred and sixty *Six* and the ninetyfirst
of the Independence of the United States of America, before me, **F. M. CROZAT,** duly commissioned and sworn
Recorder of Births and Deaths, in and for the Parish and City of Orleans personally appeared:

William Gaspar Mason a native of New
Orleans residence 83 Magazine street in
this city who by these presents declare
that

Charles N Tripp a native
of New York aged about fifty three
years died on the thirtieth instant 30th
Sept 1866 at two o'clock A.M. at the
James Hotel in this city

THUS DONE at New Orleans in the presence of the aforesaid *William Gaspar Mason*
as also in that of Messrs. *Paul E Caruth & H McGuigan*
both of this city, witnesses by me requested so to be, who have hereunto set their hands together with me after
due reading hereof, the day, month and year first above written.

R McGuigan   *P E Caruth*   *H McGuigan*

Wm Gaspar Mason

Recorder.

The Death Certificate of the
Original Oil Man of Canada.

New Orleans of congestion of the brain on the 30th of September. He was a native of Schenectady, but had resided for the past ten years in the South, where he was engaged in the examining of the mineral fields of that region. It is said he knew more practically about the mineral wealth of every Southern State than any other man; and at the time of his death he had just succeeded in making preliminary arrangement at the North, and was organizing companies to develop on a gigantic scale some of the wonderful and heretofore unknown mineral fields, which he had discovered among the oil, copper, lead, zinc and iron regions of Louisiana and Texas. His death is regarded as a public loss, and his friends, and those connected with him in business, feel most sensibly his loss as a private affliction.[6]

Two 100-acre parcels of land (Lot 13 Concession 7, and Lot 21 Concession 2) were willed by Tripp to Henry and his sisters, Ellen S. Tripp and Harriette Tripp. There was no mention of his young wife, Almira. The heirs sold the two lots four years later, receiving the grand sum of $25 Canadian dollars for each one.[7]

The man who first worked the gum beds of Enniskillen was dead. He'd had grandiose plans for the development of his asphalt beds, but the time had not been ripe and the necessary capital was missing. He saw great potential in asphalt and he sold it commercially before Williams came on the scene. Although he dreamed of asphalt to pave the roads of the world, this dream did not materialize for him. Williams picked up where he left off and developed an industry, but Charles Nelson Tripp was truly "The Original Oil Man of Canada," and, as history indicates, The Original Oil Man of North America, too.

BOOM AND BUST

In 1860, a stranger who called himself a European scientist wandered into the oil fields around Titusville. As Victor Ross in his book *Petroleum in Canada* recounts, this stranger told a group of wildcatters that he could offer better production of crude by shooting "a white-hot bolt into the bowels of the earth through an iron pipe driven to a great depth for the purpose. By the ignition of inflammable gases, thought to exist in the great cavities beneath the earth's crust, the promoter expected to produce a sufficient explosion to lay bare the subterranean reservoirs of oil. The Pennsylvanian populace, instead of viewing this proposal with the distrust and apathy traditionally accorded the first efforts of inventive minds, possessed sufficient imagination to picture the possible results, and were so convinced that the scientist minimized rather than exaggerated the feasibility of his undertaking, that they selected a small but representative committee to hang him on the spot."[1]

The stranger was, no doubt, working with nitroglycerin which had been invented in 1847 in Europe. By 1861, a year after the harsh welcome in Pennsylvania, this explosive was being used regularly to shatter rock formations deep within oil wells in an effort to assure better production.[2] However, early oil men did not welcome nitroglycerin with open arms; the people of Titusville could see the very earth shattering beneath their feet and the end of the world coming. In spite of this fear, nitro stayed, and another specialist was added to the long list of oil workers. The "well-shooter" was a member of one of the most dangerous occupations in the early oil business. As Max W. Ball observes, "a hard bump in the road or a slip at the well head, and the well-shooter could not be found or assembled for burial."[3]

Victor Ross explains the procedure for "shooting" a well:

The nitro was contained in tin tubes or shells five feet long and two inches or more in diameter, pointed at the lower end and having bail handles at the top. From five to fifteen shells were lowered with extreme delicacy to the bottom of the well, and then the 'go-devil', a five pound shell, pointed, and from ten to twelve inches long, was released downward. The shooter dashed for safety and the explosion followed.

As an example of the mental alertness and quick judgment of the well-shooters, it is recorded that upon one occasion the first tin of explosives was being lowered when the rope suddenly slackened. This could only mean that the well had unexpectedly begun to flow. The shooter realized that in a matter of seconds, six quarts of nitroglycerin would be hurled out of the well carrying death to himself and his comrades. There was no time to run; there was just one thing to do and he did it with extraordinary nerve. Standing directly over the well he grasped the shell by the bail handles as it came shooting up on the breast of the flowing oil, and although the impetus which it had gained threw him across the derrick and dislocated his shoulder, he held on and saved the crew.[4]

There is another story of a farmer who purchased an old wooden trough that a driller had once used to prepare his nitroglycerin: "He used it about the farm for months, until one day he found the boards slightly sprung. He used his axe-head to knock them into place, and they didn't find enough of him to put in a basket."[5]

Pioneer driller Bill McCutcheon used to tell another story, retold here by Selwyn P. Griffin:

McCutcheon was in charge of the drilling of a well near Oil Springs. Everything was ready for the "shot". He ran up to remove the loose planks in the upper part of the derrick for they would be blown off if they were left. As he began to move them, he saw his helper coming across the field with a pail of nitro in each hand. He called to him.

"Jim — stay where you are."

The helper did not hear; he kept coming and put the pails near the derrick. McCutcheon called to him again.

"Jim — I'm going to throw down these planks. Take those pails away out of there."

"All right," answered Jim.

McCutcheon went back to his work of loosening and raising the planks. As the first plank left his hands, he was horrified to catch sight of the two pails directly below, on the ground, where the big board would land. He watched, paralyzed. By chance, as the plank fell flat, a gust of wind caught the end of it and whirled it just enough to clear the pails.[6]

By 1864, the shooter became a very important person in Oil Springs and later in Petrolia. Nitro plants sprang up along a road appropriately named "The Blind Line." Located a mile north of the main road that ran through Petrolia, it was a road that was blown to pieces many times as plant after plant storing nitroglycerin exploded.

American drillers, who invented the torpedo to blast open rock strata, brought nitro to Canada. R. I. Bradley was the biggest operator in the Oil Springs - Petrolia area. Because of the volatile nature of nitro, railways refused to transport it from the source in Pennsylvania. Bradley was forced to bring it himself by horse and wagon, over impossible roads, to Port Huron. The ferry running the St. Clair River at this point would not let him take his load on board, and he rented a row boat and made dozens of trips, tediously rowing his loads to Sarnia. He rented a wagon and horse in the port town, and traveled over the deteriorating Plank Road to the oil field. Bradley did not live till a ripe old age, but he did die of natural causes some thirty years later when he was forty-nine.

Old timers tell a story that has drifted down through the years about a store owner who noticed a strange fluid running down the side of his building from a second story window. He ran upstairs and found a tenant working on a table near the window ledge, mixing glycerin with concentrated sulfuric and nitric acids to make nitroglycerin. She had spilled some of the mixture over the table and window sill and it had run down the side of the building. Her husband's nitro plant was located two miles away.

Because it improved the production of crude, nitroglycerin, in spite of its hazards, became big business in Oil Springs. Another method of "treating" wells was not as explosive, but just as lethal. Called "salting a well," it was practiced quietly, secretly, on the dark side of the law.

Bruce Macdonald was a twenty-five year old carpenter who'd saved his money to buy an oil well. Living in Toronto, he'd heard about the boom in Oil Springs from his uncle Charlie Macdonald who'd struck it rich there in 1862, then, with his big money, had opened a hotel in Buffalo. Four years later, Bruce wandered into Oil Springs and talked to drillers and speculators. One Saturday night he attended a burlesque show at the Oxford House Hotel, and he met a man named Campbell who said he owned a "little gold mine of a well" on half an acre of land. He hated to get rid of the well, but he would certainly sell it if the right man came along. Campbell invited the newcomer to meet him the next afternoon at the well site. In the meantime, he rounded up several friends to help him fill this dry well with crude. When Macdonald arrived, a steam engine was pumping oil into a wooden storage tank on the property. Everything looked lively and productive, and he was impressed. He paid $5000 for the well and the land. A few days after the deed was signed, the well ran dry. And Campbell had vanished.[7] There were few prosecutions for people practicing this kind of deception. The culprits vanished quickly once they caught their prey.

Salting wells was a big problem following the Civil War when Oil Springs was still booming and wildcatters began to drill deeper in the hope of finding oil gushers. Yet the greatest threat to the Oil Springs boom was not nitroglycerin or salted wells but rather the so-called Fenians. An Irish revolutionary group dedicated to ending British rule over Ireland, the Fenians became a powerful force in the U.S.A. following the Civil War. They realized that they were in a strong position to ruffle British feathers by attacking British territories in North America, and in 1866, they massed troops at various areas along the border. In April that year, they launched a raid on the New Brunswick frontier, and in June they caused alarms and excitement near Fort Erie

SHOOTING A WELL

Shooting a well with nitroglycerin at Oil Springs in 1866.

where local militia groups fought them for three days before driving them back across the border.

Seven militia companies amassed at Sarnia where a frame hotel served as a barracks. Munitions and supplies were stored near a downtown park. There were rumors that Fenians planned to cross the river here, and, by launching offensives from trains, ride triumphantly "into the heart of Southwestern Ontario."[8] This rumor spread like an epidemic, and panic swept through Oil Springs. Americans envisioned a full-scale war between British North America and the U. S. A. They decided that they did not want to be caught in the middle. The best port in this kind of storm was a home port. There was a mass exodus of Americans from the oil field.

Pumps, tank wagons, and scattered equipment stood idle in fields; three-pole derricks remained, like sentinels, guarding forgotten dreams; wild grass grew with abandon, covering the chopped-up contours of the land; stores closed. The new, grand International Hotel stood idle on the Main Street. Not a single, human client ever booked in. Eventually, mice and birds made their homes there. Gable ends blew away. Prostitutes trooped out in fancy buggies to return to Sarnia, Port Huron, and Detroit, but a few, who had a good sense for profitable business ventures, went on to Petrolia. They had been a colorful lot in their bright crinolines and parasols, painted faces and elaborate hair-dos. In the evening they had walked the board walks on the main street or had stood under the quaint street lamps, promising diverse pleasures for the right amount of gold, as

Bruce A. Macdonald recalls.[9]

Christy-Bootjack liked her log shack back in the bush, and she liked Oil Springs, so she stayed. According to pioneer Tom Evoy she gave birth to a daughter, Christy-Bootjack the Second, who continued, well into the 20th century, to work in her mother's chosen profession, with the bootjack always at the door.[10]

Although many of the pioneer hard oilers left Oil Springs, they kept their wells active there. And they never gave up hope for a return of those boom years, with some justification, for in 1862-63 thirty-five gushers came roaring in at Oil Springs. Within two years, all had ceased to produce, though, and the wells filled in with salt water.[11] Oil men reacted to the death of the great flowing wells as they would have reacted to a death in the family. They were shocked, then angry, then they denied that the era of gushers was at an end.

H. C. Crocker of Milwaukee had drilled a great flowing well in 1863, only to watch helplessly as it became choked with salt water and died a year later. He called a meeting of six local producers, including Fairbank and Vaughn. They raised $5000 to drill deeper into the 500 to 1000 foot level in search of the big oil vein. Historian Victor Lauriston reports that "heavy rigs and machinery were brought in, and three deep wells were put down by Duncan Sinclair, at that time rated among the best of the Lambton drillers. The drilling was carried to the big water vein. Five-inch pumps were put in and ran night and day. Big results were looked for. But the pumps proved unable to handle the water, and the only result of the venture was the definite knowledge that all the wells were connected."[12]

Early wildcatters had drilled too many wells too close to one another, and the natural gas pressures were bled off; wells were also left uncapped and the fields could not be repressurized. So the oil still sits at the shallow underground rock formations, for practical purposes only recoverable by the old method of pumping small amounts out at a time. In the 1970s, one oil company executive said that when his crew tried to acidize one old well to revive production, the acid came up in a dozen surrounding wells.[13]

Still, in 1881, a few of the old wells were revived by means of torpedoes of nitro; new wells were drilled from 1882 to 1886, yielding half a million barrels of crude, and since then the Oil Springs field has been in continuous operation.[14] But the great flowing wells that spewed thousands of barrels of oil into the heavens, often defying efforts at control, these spectacular gushers never returned.

THE OIL BARONS OF PETROLIA

THE PETROLIA DISCOVERY

In the spring of 1865, Petrolia had six log cabins and was described as "an undrained mud hole which extended for miles."[1] Half a dozen wells had been drilled along the banks of Bear Creek, a sluggish stream that joined the Sydenham River just beyond the town. These wells were slow producers, so slow in fact that they did not draw a stampede to the area. But that year an entrepreneur from St. Catharines, a village in the Niagara fruit belt country, came to the fledgling town. Benjamin King added the preface "Captain" to his name, but there is no evidence as to how this title was earned, whether in battle, for an honorable deed, or, like "Colonel" Drake, self applied for clout and prestige.

King was manager of North Eastern Oil Company, a refining operation in St. Catharines, in the early 1860s. He was a quiet man, not given to bragging, and, much like Hugh Nixon Shaw, he was patient and stubborn. He grew tired of the slow producing wells he was drilling on the banks of Bear Creek, and he moved a mile or two west to bush land.

A few of the American drillers in Oil Springs had migrated to Petrolia where they knew that King was having moderate success along the banks of the creek. When he moved to a location west, these hard oilers laughed at him, as they had laughed at Shaw. They believed that oil could be found only near creeks and rivers, and it was crazy to move so far away from this bonanza. Unlike Shaw, King had the necessary cash to drill without going into debt. The contract price for sinking a well was $900, but King was able to meet these high costs without borrowing money.[2] Quietly, patiently, he continued to drill down into the 300 foot level, deeper than any wells were drilled in Oil Springs. He ignored hard oilers who kept telling him that he was going too deep and in the wrong place, but he had the last laugh in the summer of 1866 when his well spewed oil into the heavens, a shock to everybody, including himself. King leaned on the expertise of hard oilers from the early boom town who showed him how to control the flow, and the well quickly leveled off at 265 barrels a day.

Drillers and speculators rushed to the site. The price of land soared. And hotel rooms were fully booked before the hotels were constructed. One man found lodging in a one-and-one-half story log shanty that already housed twenty-three other boarders and a family of seven. John Maclean, in his article "The Town That Rocked the Oil Cradle," records that "by the Fall of that year, Petrolia had 2300 residents, four churches, a school, stores, hotel."[3] But as the oil shot high, the selling price fell, as it had done after the Shaw well. Crude, which was selling at $10 a barrel in 1865, tumbled to 75¢ a barrel after the King well blew in.[4] In spite of this trend, shrewd producers began to drill in the area, and oil flooded in so quickly that refineries were unable to haul the crude away fast enough. Full wooden surface tanks covered the oil fields; built well above ground, of strong wooden slats, each tank held a maximum of 400 barrels.[5]

On a warm summer evening in August 1867, a night watchman was inspecting one of these storage tanks. He held his lantern too close to the oil where gas was rising, and there was a lightning-like explosion. As John Maclean reports, "in hours, fire was burning over 20 acres of land covered with oil tanks containing 40,000 barrels of crude, some 12 wells — one of which was flowing and burning at the same time — engines, pumps, drilling rigs and other equipment. Flames mounted hundreds of feet in the air as wooden tanks caught fire and split open. Bursting steam boilers — used to power the drill rigs — sent fragments of smokestacks, derricks and flaming debris high into the air. Flaming oil raced down ditches into Bear Creek so that one eyewitness said, 'It had the appearance of a fiery dragon raging through the valley.'"[6]

John Henry Fairbank mobilized groups of firefighters, but they were driven back by the heat. And as Victor Lauriston reports in his history of Lambton County, "rig after rig, tank after tank went up in flames and smoke. For two desolating weeks, the fire swept over ten acres of oil territory, the leaping flames illuminating the countryside, searing with their hot breath the vegetation for miles. When at last the terrible conflagration burned itself out, nothing was left on the surface except ashes; and the ground itself was charred two feet deep."[7]

Petrolia oil men realized too late that wooden surface tanks were a hazard. When they re-activated the wells and began to drill again, they did what the Oil Springs pioneers had done before them; they built storage tanks in the clay ground.

Although John Henry Fairbank continued to work his wells in Oil Springs, he bought property on Main Street in Petrolia and built a frame house there, then purchased land in the vicinity of the King site and drilled several successful wells. He also successfully speculated on land, purchasing 50 acres east of his home for $30,000, then selling this land in nine separate parcels for $58,000.[8] And, in 1864, he purchased several acres that swept majestically from a treed plateau down into a shallow valley where, at the bottom, Bear Creek flowed. The owner from Wyoming wanted $40 an acre, and Fairbank decided that the timber alone on this tract of land was worth the asking price.[9] This property eventually became the site of Fairbank's opulent estate with a grand mansion that curved majestically in turrets and wings above the creek.

Edna Fairbank realized that her husband had found security and a life-long interest in the oil industry. Although she no doubt complained about leaving her Loyalist family ties and land in Niagara, she moved, in 1866, to the pioneer town of Petrolia.

The new town of Petrolia which greeted Edna Fairbank "was not a particularly attractive site" in 1866, as Charles Whipp and Edward Phelps note in their history of the city, "with only a small and pioneer farming industry, a doubtful water supply, no

A fire burns out of control in a refinery at Petrolia in the late 1860s.

important navigable streams nor nearby seaport. It lay exactly between the logical extension of both major railroads. It was off all main lines of communication, and when the place did expand, it did not enfold the nearest highway, as is the custom of towns, but moved away from it across the banks of Bear Creek. Here, secure from outside influence and preoccupation with any other potential but its consuming interest in oil, there rose Petrolia — arrogant, whimsical, aloof, bold, brave, adventurous and contrary. It was never intended to be a town, but a place of sudden wealth, a moment in a man's life — as the California gold rush before it, or the Klondike that came after." [10]

A Toronto reporter who visited the pioneer town in 1865 nevertheless thought it was an attractive site: "Petrolia was always a better-looking place than Oil Springs, cleaner and possessing some of the elements of the picturesque. The woods in the immediate neighborhood are less tangled, there is but little swamp, and the land is rolling. A good many acres round about are under cultivation; comfortable looking farm houses stock the landscape." [11]

Shortly after the disastrous fire of 1867, Petrolia moved west across Bear Creek, where a business section grew quickly. William H. McGarvey resigned as the town's first reeve when he built a business enterprise called "The Mammoth Store" that demanded all of his time. Like his father in Wyoming, he sold everything from oil well supplies to pee-pots and spittoons. George Moncrieff, a lawyer, replaced McGarvey as reeve. At a later date, Moncrieff's sister-in-law, Charlotte Eleanor Thompson, became famous as the wife of one of the wealthy, high profile oil tycoon Jake Englehart.

Englehart was quick to expand his operations to Petrolia where, as in Oil springs, he bought crude from small producers and refined it in London. He then sent the refined product to New York where the whiskey barons shipped it overseas. But they lost money in the beginning. In some instances, "shipments to Europe turned out partially worthless, resulting in heavy reclamations," Hugh M. Grant writes.[12] Liverpool dock workers refused to handle one cargo because of the skunky smell. And in one year, his still blew skyward three times, with property damage of $8,000 and one death when "worker Oliver Odell was burned to a crisp."[13] However, Englehart cranked up his operation again, employed a dozen men and became known as "a tight-fisted, hard-driving employer."[14] Wages were low, and because he himself worked long hours, he expected his employees to do so without complaint. Other refineries sprang up quickly in London until, by the end of the 1860s, there were forty-nine big stills operating, and competition was fierce.[15]

Englehart's brother-in-law was Abram J. Dittenhoefer, a well-known New York city lawyer and supreme court judge. Through this association Englehart became friends with Cornelius Vanderbilt, widely known as Commodore Vanderbilt because of his control of the shipping on the Hudson River. In the 1860s, Vanderbilt held the controlling interest in the New York and Central Railroad, and Englehart made an exclusive deal with him to supply all fuel for the railroad.[16] This deal helped to stabilize his faltering refining business following the costly explosions; also, Englehart was a frugal young man. He lived in sparsely-furnished hotel rooms which he sometimes shared with an employee.[17]

When in Oil Springs, he often stayed at one of three hotels owned and operated by Nathaniel Boswell who was a multi-millionaire by the time he arrived in Canada. Everything Boswell touched turned to gold, although he had come to North America as a penniless immigrant. At age sixteen he left England, crammed into the hold of a sailing vessel with twelve hundred other would-be settlers bound for New York. He made his way to Toronto, and, although he knew nothing about railroads, he accepted a job to lay tracks for the Great Western Railway Company.[18] He became so good at laying tracks that he was hired to oversee the construction of railway lines in Canada, the U. S. A., and Cuba.

During the American Civil War, his sympathies lay with the South, and he purchased a river steamer called *The Silver Wave*, loaded it with cotton, sugar, molasses, and other needed supplies and ran the Yankee blockade on the Mississippi River for three years. In 1864, he was caught by Union soldiers and his cargo, worth a million dollars, was confiscated. Undaunted, he put *The Silver Wave* into service as a hospital ship, and after the war he headed for the oil boom in Canada.[19]

When he moved to Petrolia, Boswell built the American Hotel, and, as he had done in Oil Springs, he kept a livery stable of forty-five horses with stage coaches to accommodate his guests. He began wild-catting in the boom town, and hit oil every time he drilled. His first well in Petrolia, called "The Pittsburgh," rewarded him with 175 barrels a day. Crude had been selling at $2.50 a barrel, but when hard oilers like Boswell struck a number of free-flowing wells like "The Pittsburgh," the price fell to 30 cents. [20]

Fairbank tried to organize the producers into a union to be called The Petroleum Amalgamation Company, another bid to control the selling price of crude, as he had done six years before with The Canadian Oil Association in Oil Springs. Oil men, always secretive about their dealings, were skeptical about this new union, as they had been about the old one. A friend of Fairbank's wrote to him, expressing his concern: "I have just heard that you have a great consolidation scheme on foot at present, and that there is every prospect of its proving a reality this time. If it will be for the benefit of the Oil Martyrs, I sincerely hope it may be carried out, and as the matter has been explained to me, I think the company might make some money out of the affair, but again, how often have oilmen been quite sure of making money, and how often have oilmen been mistaken." [21] As Edward Phelps comments, The Petroleum Amalgamation Company "was doomed to failure because it could not succeed without the concurrence of nearly all of the oil producers. Self-interest was still stronger than self-preservation." [22]

Undaunted, Fairbank tried a year later to organize a new group that he called The Crude Oil Association. This time, the majority of producers signed up, and the organization began operations in December 1868. "As the solo agent of its members, the Association marketed, through Fairbank, fully one-third of the crude oil that was sold during the period," Phelps writes. "Because of its dominant position in the crude market, it was able to raise the price of oil, first to 75¢ a barrel, then . . . $1.25 if the refined oil was exported, or $1.62 if it were sold for local use. The price differential was established to encourage the export of oil to the United States, thereby removing it from the Canadian market. The rise in the price of oil found most of the field shut down, and tankage far from full. Speculators, sensing a profit, bought on the rising market, sending oil up to $2.25 a barrel by November 1869." [23]

Once again, a good price for crude meant that the Association was not needed, so it quietly faded away. But the men learned an important lesson with this second company of oil producers: by uniting, they wielded great economic power. It was a lesson they would not forget.

An artist's representation of Petrolia in the oil boom era.

VANTUYL AND FAIRBANK AND VAUGHN

Although John Henry Fairbank would always be in love with the oil industry, he began to involve himself in other business ventures when he moved to Petrolia. In the late 1860s, he and a partner, Benjamin VanTuyl, opened an oil well supply and hardware business. Their partnership was a hastily-written agreement in pencil on a slip of paper, but it endured with great success until VanTuyl's death thirty-five years later.[1]

Benjamin Stoddard VanTuyl was born in 1840 in New York State, the son of Thomas and Survina VanTuyl. His father was a wealthy merchant, lumberman, and landowner. At age twenty, he was hired by Eastman's Business College, Poughkeepsie, New York, to teach writing and bookkeeping. A year later, he joined the 161st New York Cavalry Unit to fight in the Civil War, and was given the rank of Major. In 1865, newly-discharged from the army, he found himself attracted by stories of the oil boom in Canada, and he journeyed northward to Petrolia. He was a man who looked every inch a cavalry officer, tall, trim, meticulously dressed, hair parted in the middle with a clipped, military moustache. He was a talented businessman, careful with money and possessing a sixth sense that told him how to differentiate between the people who were good credit risks and the people who were not.

He and Fairbank had similar dispositions and aims, and they had a long, harmonious relationship. Such was not the case, however, with Leonard Vaughn and Fairbank, who opened a private bank in Petrolia in 1869. Vaughn came to Oil Springs from Pennsylvania and followed Fairbank in drilling wells near the King wells in Petrolia. He was inclined to be a plunger, entering into business and land transactions on a whim. He was always borrowing money, usually from Fairbank, and, in 1887, when he died at age fifty-two, he was broke.[2] Fairbank and Vaughn would never have gone into the banking business together except that Vaughn had operated a private bank in Oil Springs, albeit briefly, and he was well informed on banking procedures. No doubt this side of Vaughn appealed to Fairbank.

The partners purchased an old house in Oil Springs for $70 and hauled it to Petrolia. This served as the Vaughn and Fairbank Bank, which the town's folk affectionately called "The Little Red Bank" because it was made of red barn wood. At the end of the first year, Vaughn and Fairbank realized $1.5 million in business and had a $10,000 line of credit with the Bank of Montreal in London.[3]

Many wildcatters and refiners in Petrolia climbed to success with loans from The Little Red Bank, but inside it was not always a smooth operation. Vaughn was a complainer. He often accused VanTuyl of giving too much credit to drillers. At one time the "store"

VanTuyl & Fairbank,

DEALERS IN

GENERAL HARDWARE

AND IMPORTERS OF

TUBING, CASING, PUMPS,

OIL AND SALT WELL SUPPLIES.

Shelf Hardware,
Fittings of every kind,
Paints and Oils,
Glass and Putty.

MOST COMPLETE STOCK

In Ontario at Low Down Prices.

TERMS---CASH MONTHLY.

VanTuyl & Fairbank.

The trademark for VanTuyl and Fairbank General Hardware and an advertisement from the Petrolia Topic, *29 March 1889.*

The Vaugh Block, built in 1879.

Cheque drawn on the Vaugh & Fairbank ''Little Red Bank'' by Patrick Barclay, one of Petrolia's first settlers and town treasurer for thirty years.

was in debt to the bank for $50,000. As Whipp and Phelps report, "Vaughn claimed that nearly every piece of pipe, every pump, every piece of rigging in the oil fields was being bought by the bank."[4] He must have been aware that the average yearly sales of the store amounted to $250,000, an excellent rating in those days, and that VanTuyl was a capable manager. But it appears that he was jealous of Fairbank's other partner, which did not always make for smooth sailing in the growing Fairbank empire.

Oil men often quarreled with partners, and yet partnerships were an important part of the early oil industry. Without a partner, many an early driller would not have had sufficient collateral to dig a well. Fairbank was a mediator in the Vaughn-VanTuyl problem, although VanTuyl did not let accusations escalate to open warfare, for he quietly ignored his accuser.

One oil man took a different approach. He shared a cabin and an oil well with a partner, and, after a winter in the bush he was fed up with that union. He cut the cabin in two, pulled it 200 feet away, and set up his own housekeeping quarters. "It was just too much — in the shack and on the job, too, with this wild man who chewed tobacco, spit on the floor, got drunk Saturday nights and wailed like a banshee till dawn for all the dames he claimed he'd loved and lost," he reminisced.[5]

Petrolia oil men inherited the same partnership problems that had plagued Oil Springs a few years before, and they inherited the same kind of shysters who had been drawn to the early boom town.

HARRY PRINCE

In the late 1860s and early 1870s Harry Prince was a big time operator in Petrolia, manager of the Western of Canada Oil Works and Land Company. Previous to this, he had worked as a conductor on Great Western Railway trains, and with great care had saved $30,000 with which to launch his company.

A good-looking, charismatic young man, Prince lived in style in one of the hotels that catered to the wealthy. Shortly after moving to the area, he gave dinner parties for the elite of the town. It wasn't difficult for him to get to know the elite, for early in his career he had spread the news that he was the son of

Col. John Prince, leader of militia units that had fought in the 1837 Rebellion. The elder Prince had made newspaper headlines once when an insistent reporter asked him what had happened to insurgent prisoners, and Prince replied tersely, "I ordered them to be shot, and they were shot accordingly."[1]

After launching his oil works and land company, young Prince set out to capitalize his operations in a somewhat shady manner, as Victor Lauriston reports, "by inducing a number of men prominent in Southwestern Ontario to lend him their names":

The flamboyant prospectus listed Sheriff Monroe of Elgin, Honorable Rufus Stevenson, M.P. for Kent, Honorable John Carling and Col. Walker of London, H. W. Lancy and John Henry Fairbank of Petrolia.

With this nominal backing, Prince secured extensive areas of utterly valueless, or at least unproven oil lands. . . . capitalized his company for a fabulous sum, and proceeded to sell shares to English investors.

The Petrolia oil men were probably the first to grasp the real significance of the scheme to which they had innocently lent their names, and to see the danger to legitimate oil enterprise. They openly repudiated and denounced the project, but the plausible Harry Prince was undaunted. When representatives of English investors came out to inspect his operations, he took them under his wing, kept them away from other sources of information, entertained them lavishly, showed them the wells connected for the occasion with real producers and which gushed oil under their very eyes, and so impressed them with the glories and profits of his scheme that he hoodwinked them completely, and through them, the investors they represented.[2]

Eventually, Prince skipped the country with a quarter of a million dollars, a hefty sum in those days. He lived royally in Chicago for a time, but fate was not kind to him there. He lost money on land speculations and died poor. He also left behind in Petrolia a whole series of court cases levied by the British victims, not against Prince whose whereabouts were unknown for a number of years, but against the responsible Canadians who lent their names to the company.

While tongues wagged about the Prince scandal, a mob was busy conducting a lynching. George Putnam, proprietor of the Saginaw Hotel, sold his establishment "and deposited $600 in what he considered a safe place," Victor Lauriston reports. "The money disappeared. Oil men, by a process of elimination, determined the culprit, and, taking him to the creek, strung him up to an oak tree. After a 'little' hanging, he was given a chance to disclose the hiding place of the money, but refused; three successive elevations however, induced him to change his mind."[3] One Sarnia newspaper hotly condemned this method of justice:

Supposing the party suspected had been innocent of all complicity with the offence, what a gross outrage it would have been; and for which the parties engaged in it would have had to pay sweetly, as they richly would have deserved. As it is, they should not be allowed to escape. The majority of the law ought to be vindicated; and we hope will be so; as we, at all events, do not desire the stigma to attach to our country, as it does to the neighboring republic, that the executive is so weak that there is a necessity for the mob taking the law into its own hands.[4]

Members of Canada's Secret Police Force were on the prowl through the oil fields. As early as 1864, The Western Frontier Constabulary was formed, and, in two years it was replaced by the Dominion Police Force, forerunner of the Royal Canadian Mounted Police. Pioneer members of this secret force were active in international border communities during the American Civil War. They watched and reported to authorities any unusual profit-making ventures linked to the war. There were brokers — men and a few women — "who offered a fee to Canadians who agreed to enlist in place of conscripted Americans, and then collected a bounty from the American government. . . . Some of these brokers tried to cheat recruits out of the bounty, others resorted to kidnapping in order to keep the cash coming in," Cheryl MacDonald notes in her article "Canada's Secret Police," published in *The Beaver*.[5] The oil fields were fertile ground for this kind of activity, and the private eyes were everywhere. Needless to say, they were not popular. A few were beaten up and left to die or lick their wounds. They were unarmed and therefore quite unprotected. During the Fenian raids, these sleuths checked "all arrivals and departures of the Great Western trains and those of the Grand Trunk in neighboring Point Edward. In addition, they monitored steamboats running between Port Huron and the Canadian side . . . checked activities at foundries where weapons might be manufactured, and wandered around late at night."[6]

Although it isn't known how many illegal activities were uncovered, it is known that these spies created an uncanny, nervous atmosphere in the oil patch.[7]

THE PETROLIA ASSEMBLIES

Petrolia remained, for many years, a pioneer, shanty town. As late as 1884, a young student described it this way in a school essay: "It has no buildings of any importance, they being for the most part wooden. It also has a very bad smell and visitors coming in are shocked by the smell from the refineries, the acid works and the wells. And a small, dirty creek called Bear Creek runs through it."[1]

In spite of this frontier-like atmosphere, there was big money in Petrolia, and many wealthy people wanted an upper class society with all the trimmings. A race track was built for the racing of thoroughbred horses, a Masonic Lodge united in brotherhood some of the high profile oil men including Vaughn and Englehart, and, as early as 1867, VanTuyl and Moncrieff were members of a committee that inaugurated Petrolia Assemblies. These balls, held two or three times a year, gathered together the so-called elite of the town. Professional harp musicians were brought from London and a well-known orchestra made up entirely of Negroes often came from Detroit. The Assemblies were held in the Opera House (sometimes called The Town Hall because it was used for town gatherings of various kinds), and later in the ballroom of the Fairbank's palatial home when it was finished in 1890. Helen Corey, a pioneer, when describing the Assemblies, became excited while remembering them, and exclaimed: "Oh the music, the parties we had up there were simply beautiful. There wasn't anything in New York that could touch them."[2]

The events called for chandeliered rooms, champagne, caviar, and hors d'oeuvres; the men wore dark suits, cutaway collars, bow ties. The women's gowns were a colorful array of fashion — woven satin with fitted bodices, scoop necks, cap sleeves, wide crinoline skirts worn over hoops. In the beginning, people danced the quadrille, the famous contredanse that originated in France in 1700. Later dances became more risque with the daring waltz and the two-step.

By contrast, prostitutes livened up the area by offering exciting diversions to those hard oilers who could cross their palms with gold. One "Madam" was known as Godiva by her regular customers. She was a colorful, gypsy-like redhead of thirty, who often threatened to mimic the original Lady Godiva by riding nude on horseback through the center of town, to shock the "excrement out of religious fanatics," as hard oiler Bruce A. Macdonald has stated so eloquently and decorously.[3] It isn't known whether she actually carried out this threat. Petrolia seemed to like the title "Ladies" for there was another "Madam" who called herself Lady Adair. According to Arnold Thompson, another pioneer, she lived and conducted her trade on the main street of the town.[4]

When the opera house was finished and ready to present live performances, the local Baptist clergyman ranted and raved at the sinfulness of the theater. He told the town council that he did not want the good, virtuous members of his congregation saddled with property taxes to pay for stage props, curtains, and other trappings guaranteed to send souls straight to hell.[5]

About this time, the oil business took a bizarre turn when one of the hard oilers developed a gas that he said would turn dead bodies to stone. Under the headlines, "Preserving Your Friends," an article in the *Sarnia Observer Advertiser* read:

The embalming business is threatened to be superseded by a new process for preserving dead bodies. An antiseptic gas has been discovered which, when introduced into the burial case — from which the atmosphere is exhausted — the cadaver absorbs, and in the course of ninety days, becomes as hard as stone. The experiment was made on the body of a drowned person and the corpse is now exposed to the air without the slightest indication of decay.

The cemeteries in the suburbs are gradually filling up, and it may be that objects of bereavement will, in time, become articles of the household.[6]

On a broader issue, Canadian oil was doing well on the foreign market. From 1869 to 1874, two-thirds of the total output was exported. "As a consequence, the industry assumed a dual structure, consisting of large exporters and a number of smaller refineries supplying the domestic market," Hugh M. Grant notes.[7] The oil industry, however, was ever at the mercy of people who saw an opportunity to make a fast buck. Englehart and Company in London saw this opportunity in 1869. In association with Judge Ebenezer Higgins of Chicago, the company leased fifty-two refineries in the London-Petrolia area, then closed them down. "They purchased all stocks on hand at 18¢ per gallon and all the crude offered," Edward Phelps reports. "They then waited for the market price to rise to a point where it was profitable to begin refining, when they [would] dispose of their supplies. The public raged over this 'corner of the oil market'."[8]

The ploy was really a bid by Higgins and Englehart to dominate the entire oil trade in Canada, but Canadian producers and small refiners hurried to organize to maintain control. The first to emerge was The Carbon Oil Company, initiated by James Miller Williams, in partnership with E. B. Parsons, a Toronto wholesaler; the company was financially backed by Fairbank. A huge oil-producing, refining, and manufacturing concern, the company had a big still in Petrolia, a vast underground storage area, and a pipeline network to every corner of the oil field. Six months after it was formed, the company reported in the *Sarnia Observer Advertiser* that "70,000 barrels of oil had been delivered and 67,000 barrels had been sold — 31,322 barrels at 75¢ and the balance at prices varying from $1 to $1.62."[9] Carbon Oil was off to a good start.

The Petrolia Assemblies were the social event of the age.

In 1871, Williams left the company and threw in his lot with William Cochrane of Petrolia to operate a new refining and treatment plant in town. With so much opposition, the Higgins-Englehart partnership collapsed, and Englehart thought it wise to become allied with a Canadian firm; he formed a company with the Waterman Brothers to refine crude in London. "Herman and Isaac Waterman arrived in Canada in 1855," Hugh M. Grant writes, "and had been active in the oil refining business since 1861. They described themselves as 'German Israelites'."[10] These three firms — The Carbon Oil Company; Williams and Cochrane; Waterman, Englehart and Company—"accounted for over 40% of the total oil production in 1870-71. They also purchased the output of several smaller refineries for treatment and export."[11]

Following his unsuccessful attempt to dominate the industry with Higgins, Englehart continued to live and operate his business in London, but he spent much time in Petrolia where he carefully monitored production and refining operations. At the tender age of twenty-three, he began to emerge as a strong force in the industry. He often traveled to New York where he consulted with his partners, the notorious whiskey barons, Sonneborn, Dryfoos and Co., who took care of his exports. Sometimes, he dined at Delmonica's, the fashionable New York dining club patronized by the wealthy and famous. He spoke well, with a certain eloquence. He dressed in dark suits, tailor-made to fit him. Although only about five-foot nine inches tall, he seemed taller because he stood straight with confidence. Like a dashing cavalier, he always wore a fresh flower in his buttonhole. As he grew older, he cultivated a Vandyke beard.

He had influential relatives in New York — a brother-in-law, Dittenhoefer, the judge, and a brother, I. Albert Englehart, a lawyer associated with Dittenhoefer's law firm. He began to build a roster of famous friends, including Cornelius Vanderbilt, the shipping and railroad magnate, and Isaac Guggenheim, whose family became wealthy through mining ventures in the American West. But Petrolia would give him an opportunity to become rich and famous, a man with tremendous power and influence, before he turned forty.

CORPORATE VENTURES

Jake Englehart was surprised in 1873 by an ambitious scheme in Petrolia, the formation of The Home Oil Company, a producers' association led by Fairbank to check the escalating power and control of the refiners in London. Fifteen shareholders built a huge still costing $37,000, and the company's capital was set at $50,000.[1] By 1874, this new plant was able to process 3,000 barrels of crude a week. All original shareholders, except for James Miller Williams, were residents of Petrolia.

By this time, James Miller Williams had moved his refining operations to Hamilton. And by the time that the Fenian raids cleaned out Oil Springs, he was, with great marketing skills, selling his kerosene to Europe, China, and the United States. With oil money, he built an elaborate Victorian home in a wooded area of Hamilton.

Ever keen on politics, Williams followed with great interest the progress being made by the various factions of British North America towards the birth of a nation. A Great Coalition had been formed as early as 1864 to seek general federation. A conference that year at Charlottetown, Prince Edward Island, with representatives from the various British North American colonies, pointed out the necessity for union. The American Civil War and subsequent Fenian Raids had placed the colonies in a nervous, awkward position. Great Britain had sent troops across an ocean to help defend these territories, but the "Mother Country" was unhappy with this expense. Early in 1867, the British North America Act became law and it set up the terms of union for the British colonies as the Dominion of Canada.

Williams was the first of the hard oilers to seek election in the new Dominion. On a Liberal ticket, he ran as Hamilton's provincial representative to the Ontario legislature and was elected. Records of his term of office indicate that he was a man of few words: "His speeches were rare and brief, and restricted to the interests of his constituency."[2]

Williams turned forty-nine that year, and he began to drift away from the oil fields, leaving his son, Charles, in charge of production and refining operations while he formed a company to manufacture brass stampings and open a private bank and insurance firm.

Throughout 1874, oil prices continued very low as producers and refiners vied for control. Early in the year, when production had fallen to 6,000 barrels per week, the Petrolians attempted to seize the upper hand. As *The Monetary Times* reported, "It is finally decided that the production in this place will, if possible, be made up into marketable lots here, thus giving ample employment for all the refiners and The Home Oil Works. London refiners will have to beg for crude and pay the price asked."[3] The Home Oil Company would endure for eight years, longer than any other producer/refiner consortium up to that time, and would be a thorn in the side of its enemies outside of Petrolia. The shareholders were well aware of what could happen when companies permitted outside involvement.

The Carbon Oil Company, organized in 1869 by Williams, was a good example. A gigantic oil-producing, refining, and manufacturing company with a big still in Petrolia, Carbon Oil began to flounder four years after conception. The plant had caught fire three times, totaling $20,000 worth of damages, but fire did not bring about the collapse of Carbon Oil. Early in its career, the company hired the services of Soloman Sonneborn in New York to export its oil. This was the same Sonneborn who was affiliated with Sonneborn, Dryfoos and Co., the whiskey rectifiers who, seven years earlier, had, with Englehart, turned to oil to legitimize their trade. Jay Cooke, a railroad financier in New York, went broke in 1873, and Sonneborn, affiliated with Cooke, found himself in need of money. He sued Carbon Oil Company for $100,000 that they owed him. "Desperately, Fairbank tried to rally the company's creditors to keep afloat, but the company went under late in 1873," Charles Whipp and Edward Phelps recount. "The Vaughn and Fairbank bank lost most of the $30,000 it had invested, but the most crushing blow fell on oil producer John D. Noble [the ship-owner who, eleven years before, sold his fleet to join the Oil Springs stampede]; he had advanced Carbon Oil $10,000 and was forced to deed off his oil properties to cover his losses."[4]

Englehart, too, was affected by these problems. The R. G. Dun & Company Collection in New York noted that, in 1873, Englehart's firm had "only been moderately successful and they had to get assistance from their friends."[5] A few months after Carbon Oil collapsed, Englehart severed connections with Sonneborn, Dryfoos and Co., and hired Jonas Sonneborn to be his export agent in New York. Jonas was Soloman's uncle, and was said "to be worth in excess of one million dollars. He had been the financial strength [behind Sonneborn, Dryfoos and Co.] underwriting the various activities of the firm."[6]

The following year, Englehart became a British subject, pulling away from his American roots.[7] He organized The London Oil Refining Company, an amalgamation of several oil refineries in London, "who closed down most of the capacity and restricted production and marketing. This effort, which achieved some success, broke up after 15 months. There was too much independent capacity outside their control as well as imports to fill shortages when they appeared," Robert Page concludes in his study "The Early History of the Canadian Oil Industry, 1860–1900."[8] The Home Oil Company in Petrolia supplied much of the independent capacity, and stood secure when the London amalgamation broke up.

Englehart was emerging as an organizational genius, however. Although he failed in his attempt to hold together The London Oil Refining Company permanently, it was a step towards control of a wayward group of stillmen who preferred to work alone, in secret. It harnessed, for just over a year, the most important refiners in London: Frederick Ardiel Fitzgerald, born in 1840, one of the builders of London's waterworks who had become interested in production and refining in 1865 with plants in Petrolia and London; John Minhinnick, a plumber-turned refiner; Thomas Hodgins, a former carriage and wagon maker; his brother, Edward, who manufactured wooden oil barrels; and John Geary, a lawyer-turned refiner. These men would appear later in a more sophisticated organization called Imperial Oil.

An enmity, distrust, and downright hatred existed between Petrolia producers and London refiners. In 1866, a railroad spur line was built from Petrolia to the Great Western terminal at Wyoming, and eight years later it has been reported that the "railway began to charge more to haul crude out of the town than refined oil. There is one report that The London Refiners prevailed on the Great Western to institute this discriminatory rate structure to hit back at The Home Oil Company."[9] The episode drew a parallel with Rockefeller's underhand treatment of producers in the U.S.A. a few years before, and Petrolia oil men screamed loud and long for a competing railroad which was not built for five years.

Producers and refiners called a truce when they dealt with the government. As Hugh M. Grant notes, early oil men found it easy to drill at Oil Springs and Petrolia because "government regulations were virtually absent, since mineral rights had been alienated with the sale of title to the surface, no crown royalties were paid, there were no laws regarding well spacing, and the rule of capture prevailed."[10] This rule of capture meant that oil men could sink a well and claim as their own the oil they brought to the surface, even if this oil seeped into their well from an underground, common pool. The rule of capture prevailed, not only at Oil Springs and Petrolia, but also at Titusville, Pennsylvania. It caused fierce competition among producers, as well as heated arguments, fist fights, and death threats.

However, early in the boom years, the government offered some control to help oil men. "Producers and refiners had always combined to seek tariff protection for their economic interests," Robert Page explains. "In 1868 duties were placed on oil, ranging from six to fifteen cents for crude and refined products; the duties provoked protests especially in

A drilling crew stands beside a well in Petrolia in the 1880s.

the Maritimes which relied on American products."[11] When Prime Minister Alexander Mackenzie came to power in 1873, he was faced with a dilemma: "As the member for the oil region, he had promised adequate protection; while as Liberal leader he had many caucus members who wanted significant reduction in tariffs such as oil. In 1877, Mackenzie revised the duty to make it a uniform six cents per wine gallon. The oil interests protested loudly to their member, but to no avail. In the election the following year they strongly supported John A. Macdonald's Conservatives and their premise of increased tariff protection . . . . Once back in power, the Conservatives chose not to increase the protective tariffs."[12]

Tariffs were all important to the Canadian industry, for they put more emphasis on Canadian oil and gave producers and refiners an advantage over their competitors in the U.S.A. Even at that, American oil did come into Canada. It was more expensive by at least six cents a wine gallon, but many consumers liked it better than the Canadian product. James Miller Williams' wife, Melinda, purchased Pennsylvania kerosene for her lamps because it didn't smoke and didn't smell of sulphur. However, she firmly advised her grandchildren that they were not to tell their grandfather about her traitorous ways. And they did not tell him.[13]

PETROLIA FAMILY POLITICS

The private lives of the oil men were often more turbulent than their public lives. Although John Henry Fairbank built a comfortable frame home for Edna, and wealth gave her power and status, she was a depressed, unhappy woman because she imagined herself ill much of the time. She was, in fact, a hypochondriac. In her estimation, Petrolia was a backward, frontier town that reeked of oil. She was homesick for the Niagara region that she had been forced to leave in order to be with her husband. Niagara was a well-settled, well laid-out area with a railroad that had brought civilization right to her door. She was ever aware of her Loyalist family there, and the prestige given them, even though they were far less wealthy than her husband.

Edna found comfort in religion. She was confirmed in the Anglican church in Rochester, New York in 1866 while convalescing from illness. A year later she gave birth to a son, Frank Irving Fairbank, who died in August 1867. Another child, Huron Hope Fairbank, was born in 1868, but lived only two months. Edna grieved deeply over these two deaths and returned to Niagara for a two months' rest at the home of her parents. In 1869, she gave birth to a daughter, Mary Edna, nicknamed "May", who was healthy and strong and lived a long life. When May was born, Fairbank's oldest son, Henry, was thirteen years old, and Charles was eleven. The boys were students at the London Collegiate Institute, the predecessor of Helmuth College. Upon graduation from the Collegiate in 1875, Henry attended the University of Toronto to study science for four years in preparation for medicine. A few years later, Charles was one of the original eighteen cadets who entered the Royal Military College at Kingston.[1]

When May was four years old, Edna, ever searching for cures for her imaginary illnesses, took the child to health resorts in England and Ireland.[2] This was the beginning of mother-daughter journeys to find health and happiness that seemed to elude Edna all of her life. Later, May, too, became a complainer and a hypochondriac, like her mother, Charles O. Fairbank reports.[3]

By the mid-1870s, John Henry Fairbank was one of the wealthiest men in Petrolia, and with this wealth new problems emerged. Distant cousins plagued him for money; a compassionate man, he remembered his own debt-ridden days and often sent gifts of money to these "poor" relations. Word has drifted down through the years that Fairbank was a ladies' man. He liked a pretty face, a well-turned ankle, a seductive figure, and women liked him. His wealth gave him power and status; he also dressed well, talked and walked with confidence, and exuded a quiet charm. However, Edna was a jealous woman. When she and her husband appeared together in public, she walked beside him and always carried a black umbrella. If he paused to admire or nod at a pretty woman, she admonished him by a quick

John Henry Fairbank and his wife, Edna, in the 1880s.

> Petrolia 21st January 1887
>
> My Dear Sir
>
> The Dominion Elections are now fixed for Tuesday 22nd February
>
> I am again the Reform Candidate for a seat in the House of Commons as representative for East Lambton
>
> If my course in Parliament during the past few years and my business record for many years has merited your confidence I trust you will give me your vote and hearty support —
>
> Faithfully Yours
>
> JH Fairbanks

*John Henry Fairbank's open
letter to voters, 1887.*

tap on the head with the umbrella.

Fairbank was elected Member of Parliament for East Lambton and he had also served one term as Mayor of Petrolia. Jake Englehart, however, was never interested in signing up for a political position, whether it be municipal, provincial, or federal. Knowing that a politician's life was open to public scrutiny, perhaps he avoided politics because he did not want it known that he was affiliated with whiskey distillers in the United States who were being chased by the government for violating revenue laws. Englehart would have been a good, dedicated politician. He was tuned in to what people wanted, especially in the oil business, and yet, as a refiner, he did not always see eye to eye with producers, a trait that certainly would not have helped him at the polls. Because he was a Jew, people did not always trust him, even though he became a devout Anglican. In Petrolia, he not only attended Christ Anglican church, but gave a gift of eleven chiming bells (some of the finest in Canada) to the church.

Throughout the 1870s and 1880s, Englehart lived a bachelor's life, boarding in hotels and working sixteen hours a day. Like Fairbank, he was a compassionate man. Once, he saw an employee in an off-limits' area of his refinery. In a burst of temper, he fired the man. Two weeks later, he re-hired him and paid him back wages for the time he had missed.

Early in his refining ventures, Englehart had become interested in railroads, for he knew how important rail transport was to both producers and refiners. In 1872, he went to a railroad convention in

Saratoga, New York, where he met and made life-long friends of high profile railroad people.[4] His knowledge of railroads and the politics involved in the business caused him grief when the Petrolia producers accused him and the London Refiners of tampering with tariffs on oil coming out of Petrolia. But his knowledge of railroads led him to a rewarding second career at a later date.

Benjamin VanTuyl and his wife, Kate, had three sons, born in Petrolia in the 1870s. Later, two graduated from the Royal Military College in Kingston, and a third, Benjamin, stayed on in Petrolia and manufactured explosives. Apart from long days spent in the hardware store, VanTuyl was also in business for himself drilling artesian wells, and he operated oil wells of his own.

In the 1870s and 1880s, typhoid fever was on the rampage in Petrolia. In one year there were eleven deaths from this disease, but other diseases also took a toll in young lives, like dysentery.[5] Kate VanTuyl died of dysentery at age forty-three. Several years later, VanTuyl married Emma Hovey Williams of Michigan.

Although he had fought with distinction as a cavalry officer in the American Civil War, VanTuyl was a pacifist. He belonged to the fraternal organization called Knights of Pythias, incorporated by an Act of Congress on 5 August 1870. Founded in 1864, the Knights struggled to heal the emotional wounds and hate caused by the war, and to offer benevolence and friendship to all peoples. VanTuyl was Vice-Grand Chancellor of the Knights for Canada.

Leonard Vaughn's grandfather had been a doctor, and Vaughn, who was interested in herbal medicine, often gave advice on health matters to patrons of the Little Red Bank. Apart from his work at the bank, he owned producing wells in Petrolia and Oil Springs. He and his wife, Sarah, had three children — two girls and a boy. In 1871, Vaughn became treasurer for the village of Petrolia, accepting the position after Hugh Smiley left in disgrace, owing the village $1,300.[6] This was the same Smiley who, nine years earlier, had helped Shaw drill his famous gusher. Although Vaughn handled money well for other people, he had a difficult time managing his own personal finances. He over-extended himself on land speculations, and, in 1887, at age fifty-two, he died broke. Fairbank is reported to have paid off his debts, including funeral costs.

George Moncrieff, who presented himself in the *Petrolia Advertiser* as a "barrister, attorney-at-law, Solicitor in Chancery and Notary Public," was married in 1873 to Isabella Thompson, daughter of Thomas and Ellen Thompson, a prosperous farm couple from Adelaide County near London.[7] Their first child, George, was born two years later, and after that, four more children arrived about a year apart. Isabella's sister — Charlotte Eleanor Thompson (nick-named "Minnie")—lived with the Moncrieffs.

During the late 1870s and early 1880s, the population in Petrolia began to divide. Wealthy oil men built lavish Victorian homes in a secluded area called Crescent Park. Here, on a picturesquely-curved street, spacious brick and frame houses rose in splendor,

George Moncrieff served as Mayor of Petrolia in 1874 and went on to become a popular Member of Parliament.

rivaling the most elite districts of New York or London, England. The houses looked out on a circular park of grass and shrubs where children could play. The Moncrieffs moved into one of these houses, an elegant home with brick fireplaces, large living room, oak dining room, sun room, five bedrooms, and servants' quarters. The back part of the property sloped down to a shallow valley where Bear Creek flowed.

Charlotte Eleanor Thompson had joined the family when she was ten years old, and she attended school in Petrolia. In her sister's home, she was surrounded by luxury and a family who were devoted to one another and to her. Here, she met and dined with famous oil men like Jake Englehart, and later, when Moncrieff was elected to parliament representing West Lambton, she met leading politicians of the day, including Sir John A. Macdonald, the first prime minister of Canada. Queen Victoria bestowed upon Moncrieff the honorary title "Queen's Counsel" awarded to lawyers who were thought to be worthy of arguing cases for the crown.

Crescent Park residents often went to Detroit for weekend buying sprees, and many had resort homes along Lake Huron and Georgian Bay, in Florida and California. One Crescent Park resident told a reporter, "This is the snooty end of town."[8] Fairbank, however, did not live in this area. An independent man, not given to following the pack, Fairbank and his family lived in the downtown area, close to the Little Red Bank. People were jealous of him and his successes, "and he was sometimes regarded with suspicion and

A work crew moving a drilling rig with horse power in Petrolia in the 1870s.

An original three-pole derrick in the foreground with a Canadian Pole Drilling Rig behind on the grounds of the Oil Museum of Canada.

publically maligned," Whipp and Phelps note. "A. C. Edward, who became mayor of the town, once attacked him so vehemently that Fairbank was impelled to bring a libel and slander suit against the man. The case was dropped when Edward apologized."[9]

In 1867–68, William H. McGarvey became reeve of the village for the second time, and that year he married Helena J. Wesolowska of Mount Clemens, Michigan. A year later their daughter, Nellie, was born, then a son, Frederick, and, in 1876, another daughter, Mary (nick-named "May") arrived.

McGarvey owned several wells in Oil Springs and Petrolia, and was part owner of the famous "Deluge" well that, in 1873, leveled off at 600 barrels a day. He was also involved in refining, and, in 1880, was reported to have a refining operation worth $5,000.[10] In 1874, when Petrolia became a town, McGarvey celebrated his thirtieth birthday. He was no longer the lean teenager who had fallen in love with oil at the boom in Oil Springs in the early 1860s. He had put on weight and was broad-shouldered and stocky. He had grown a thick moustache that drooped slightly at the corners, giving him a dour look. Charlie Whipp once commented that McGarvey "came from a family of salesmen, and was very persuasive and convinced of the oil economy. He could sell himself, and sell oil; he was a dynamic figure. . . persuasive and dynamic."[11]

McGarvey was instrumental in making the Canadian Pole Drilling Rig famous. These rigs were equipped with runners, and much like sleds, were moved from well to well. In her study of *Technology in Ontario* , Dianne Newell reports that as early as 1866.

Using a rotary drilling technique solid wooden rods was introduced to the Ontario oilfields at Petrolia to replace cable-drilling (the old spring-pole method). Lengths of hardwood were screwed to each other during the course of drilling, being lowered into or withdrawn from the hole by means of a tall tripod erected over the well. The use of poles had a long history in drilling artesian wells since the poles could hold drilling tools together better than a rope or cable while boring through difficult rock and bore straighter holes. . . . The man who introduced this pole technique to the Ontario petroleum industry was a local businessman by the name of William H. McGarvey. In brief, the components of McGarvey's system, the wooden rods, down hole tools, locomotive-type boiler, and portable steam engine driving a band wheel, with crank and pitman to drive the walking beam and hoisting drum, were largely conventional. The innovative feature lay in the particular pattern of their combination and in small design details, such as the wing guide and reamer, which McGarvey added to them. . . . Its improved version was known as 'The Canadian System' of drilling. The cost of a complete Canadian drilling rig, manufactured at Petrolia in 1890, was a modest $1,700.[12]

In the 1870s, McGarvey, like other wealthy oil men, built an elegant Victorian home in Petrolia. Although it was not situated in the Crescent Park area, it was built along the same lines as many of the houses there, in Victorian style with turrets and towers and wings, crafted in brick and hardwood. These oil men were the *nouveau riche* and lived a lifestyle accordingly, but the town had trouble keeping pace, as Whipp and Phelps observe: "While business continued to build. . . there was comparatively little municipal government activity. The old wooden sidewalks were in deplorable condition and there were numerous cases of broken arms and skinned shins as unwary pedestrians plunged through the rotting boards. In 1877, so many buggy wheels had been snapped off on the main street that it was re-planked through the flats and gravelled in the west end." [13]

Canada's second governor-general, Lord Dufferin, and his wife, the Marchioness of Dufferin, visited Petrolia at this time, and she, in her daily journal, gave a sketch of the oil patch: "We started at eleven, with a large party 'on board the cars,' to visit the oil-wells of 'Petrolia,' where we saw the oil as it comes up through the pump — thick, black, and mixed with water. We also saw the process of looking for a well, 'sinking a shaft,' and all the machinery used. The oil leaves Petrolia free from water, but black and thick: the refining is done at London. The oil district is, of course, ugly, the ground black and swampy. Stumps of trees and wooden erections — some like enormous barrels — cover the whole place." [14]

Oil fever was rampant in Petrolia. People lived, slept and dreamt oil. In 1871 there were forty-two refineries churning out nearly three million dollars worth of illuminating oil. [15] And oil magnates felt that the best years were yet to come, at home and abroad in foreign oil fields.

FOREIGN DRILLERS AND IMPERIAL DREAMS

In the summer of 1874 a crowd of people packed the railway station in Petrolia. Local bandsmen thumped out a lively rhythm, and just before the train departed they broke into the nostalgic strains of "Will Ye No Come Back Again." Three hard oilers waved goodbye from the windows of the moving locomotive. Driller Joshua Porter, engineer Malcolm Scott, and scaffoldman William Covert were the first of the Enniskillen drilling crews to "go foreign," a trend that continued for the next sixty years. They came back again — and again — and again — with exciting news of lands on the other side of the earth. After six months, Covert wrote that he lived in a six-room bamboo house in the hills of East Java, he had servants at ten cents a day, the men had built a rig, and they were busy erecting a refinery. [1]

Drillers left Enniskillen because work at home had been affected by a recession, Whipp and Phelps explain: "All of Canada suffered a depression in this period. Gorged on the financial excesses and

*The Roll of Foreign Drillers
at the Oil Museum of Canada.*

corruption of the post-civil war period, the U. S. underwent a business panic in 1873 and the depression that followed was reflected in Canada."[2] The depression lasted ten years, and in that time, hundreds of hard oilers left Enniskillen to wildcat in scattered parts of the world. "They usually worked under the most hazardous conditions. In temperatures in excess of 120 degrees they drilled the first producing well in the Persian fields. The late Fred Edward recalled drilling in the baking heat of a Persian valley while British troops fired from the hilltops at marauding Arabs. Dysentery and malaria were common, and violent deaths not unknown. . . a Petrolia boy on his first trip to Sumatra died of a knife in the back. Foreign drillers of long experience never, never walked through a door without first looking to see who might be standing behind it."[3]

In spite of the recession and the exodus of highly-trained men, Petrolia remained calm and business struggled on. That year refiners shipped out more than seven million barrels of oil to foreign ports.[4]

Meanwhile in London, Englehart and Company re-organized following the demise of the London Oil Refining Company, and the new firm teamed up with Isaac Guggenheim, the twenty-two-year-old eldest son of the famous Guggenheim family that had amassed wealth through mining ventures in the American West. Isaac, who married Carrie Sonneborn, youngest daughter of Jonas Sonneborn, was reported to have put $25,000 into Englehart and Company, an injection of capital that was sorely needed by the firm.[5] Rumors floated around Petrolia that young

Guggenheim, no doubt financed by his father, purchased a steamship docked at Sarnia that was used by Englehart to transport oil straight to overseas' ports.

In the fall of 1877, following the breakup of the London Oil Refining Company, there was disunity among refiners; producers in Petrolia took advantage of this mayhem to organize a new combination, the Mutual Oil Association. Englehart was one of the founders of this operation, throwing in his lot with producers, for a change of pace. This collective effort involved 154 members and was able to raise the price of crude oil from $1.10 a barrel to $2.00.[6]

A year later another railroad, the Canadian Southern, pushed its line into Petrolia, supplying stiff competition for the Great Western that had angered producers by raising rates a few years earlier. Business boomed. "New buildings rose and drilling of new wells resumed, while abandoned wells were restored to production," Whipp and Phelps recount. "Vaughn built the Vaughn Block. . . the first major brick business block in town." However, in January 1879, "the price of crude suddenly dipped to $1.70 and it became apparent that the Mutual Oil Association could not hold back much longer the huge stock of oil (300,000 barrels) in storage. . . . Oil continued going down and by May it touched a dollar a barrel. One day that month, which the *Petrolia Advertiser* dubbed 'Black Friday,' oil hit 40 cents a barrel. The Mutual Oil Association collapsed."[7] Even before this demise, refiners made "secret bargains with renegades," undermining the association's aim to hold back oil until the price was forced up.[8]

Two Petrolia foreign drillers at work in Borneo in the early 1900s.

John D. Rockefeller once observed that "loosely organized associations were but 'ropes of sand'," and this was true of the London Oil Refining Company and the Mutual Oil Association; both "were unable to maintain discipline among their members and their ability to regulate output was severely impaired," Hugh Grant observes.[9] Englehart, however, had set his sites on a mammoth refining operation, and setbacks like the collapse of Mutual Oil did not deter him from reaching his goals. He sold one of his two London refineries.[10] Five years earlier, this refinery was reported to have employed fifty men with a yearly output valued at half a million dollars.[11] With this sale, Englehart became a millionaire before the age of thirty.[12] Immediately, he cranked up another operation and called it the Silver Star Refinery.

In 1878, he asked Petrolia to give him five tax free years for his big still if he moved the operation to the town. The *Petrolia Advertiser*, in an editorial, backed this proposal:

We refer to Messrs Englehart's Oil Refinery... these gentlemen ask for five years' freedom from taxation. We say, grant it them.... From what we know of Messrs Englehart's present operations and from what we believe is the future extent of them, there is no manufacturing firm in the length and breadth of Canada that we would sooner have amongst us.... Let us be frankly affirmative, remit the taxes as required of us and extend the right hand of fellowship in every possible way within

our corporate power. We have no hesitation in predicting that such a refinery as the one we refer to will employ the equal of one half of our present population by-and-bye.[13]

A year later Englehart moved his Silver Star operation to Petrolia where, a year earlier, he had bought out the old, forsaken Carbon Oil Works with its big still, underground storage facilities, and pipeline networks. Here, he set up his new facility.

Englehart was certainly not new to Petrolia. He had lived there, often for months at a time, in local hotels while he purchased crude at rock bottom prices for his London refinery. It is believed that at one time he had cooked up a special deal with the Great Western Railway to ship out his crude at prices far below the going rate. Englehart was persuasive, and in thirteen short years had learned the many cut-throat tactics of the growing oil business. He would always be a survivor.

In February 1879, he ran an ad in the Petrolia paper:

SILVER STAR REFINING COMPANY, Having Removed their works to Petrolia, Ontario are now prepared to furnish their brands of Illuminating and Lubricating OILS.... We claim superiority in every way for our products. We Ask a Trial And will guarantee satisfaction. We do not claim a cheap Oil, but we do claim THE BEST. J. L. Englehart & Co. — and — shippers of Petroleum.[14]

A new, exciting era was dawning for the Canadian oil industry, and Englehart would be very much a part of it.

On the other side of the Atlantic, a pioneer Canadian geologist died: Sir William Edmond Logan had retired ten years before, and had gone to Wales where he lived with his widowed sister, Elizabeth Logan Gower who had inherited Malgwyn Castle from her wealthy husband, Abel Lewes Gower. In 1875, Logan, age seventy-seven, died at the castle.

As founder and first director of the Geological Survey of Canada, he had been an eminent geologist of world renown; he had encouraged Charles Nelson Tripp to develop his asphalt operation in Enniskillen and he gave him a world audience at the International Exhibition in Paris, France in 1855. A eulogy, delivered to the Natural History Society of Montreal shortly after his death, summed up Logan's contribution: "No man has done as much to bring Canada before the notice of the outside world and no man is more deserving of being held in remembrance by the people. Just as statesmen and generals have risen up at the moment of greatest need to frame laws or fight battles for their countries, so Sir William Logan appeared to reveal to us the hidden treasures of nature just at a time when Canada needed to know her wealth in order to appreciate her greatness."[15]

Logan once described himself picturesquely as follows: "I fancy I cut the nearest resemblance to a scare crow. What with hair matted with spruce gum, a beard three months old... a pair of cracked spectacles... a waistcoat with patches on the left pocket where some sulphuric acid, which I carry in a small vial to try for the presence of lime in the rocks, had leaked through."[16] He was buried in the churchyard at Cilgerran, South Wales.

FOUNDING IMPERIAL OIL COMPANY, LIMITED

Jake Englehart never undertook a project unless he felt it had great potential to be a successful enterprise and a money-maker. Once this was established, he threw his heart and soul into it. The Silver Star Refinery was such a venture. Within a year he improved and updated the existing facility until it was worth well over a quarter of a million dollars. It stood on a 50-acre spread of land between two railway terminals in Petrolia, and it was rated the largest refinery in Canada. An early Canadian Atlas described it as follows:

An elaborate steam pumping system drew the oil direct from nearby wells or from storage tanks into immense underground reservoirs with a capacity of about 100,000 barrels.

From these the oil was drawn into six tubular iron stills, each of 350 barrels capacity, fired by gas from the wells. The resulting "distillat" was forced into an agitator which treated around 1800 barrels at one time. Run into three vast settling tanks, the oil was drawn off through pipes to the barreling and shipping shed. The barrels were made in the company's own cooper shops, though foreign oils were shipped in 10 gallon tin cans, also made at the plant. . . . The product was loaded into cars, the foreign going to India, China, Japan, Australia and South America without breaking bulk except once, at New York harbor.

Steam was generated in four large tubular boilers, and the engine house was equipped with eight steam engines and force pumps, aggregating, nominally, 150 h.p., but capable of doubling that. The plant had its own water works system, with 12 hydrants operated by an independent engine which also pumped water from a well half a mile distant into a 5000 barrel tank at the works. The refinery had its own fire brigade, fully equipped.[1]

Englehart also retained a small, London refinery under Englehart and Company, which was worth $75,000.[2] These London and Petrolia refineries would be the nuclei of a big corporation in the next few years.

But Petrolia was not the only oil producing region in Canada at that time. Oil Springs was still churning out its quota of kerosene and related products. John Henry Fairbank still had one hundred producing wells there, and in 1881 the town would experience an exciting revival of those early

halcyon years, although this second boom, like the first one, would not last.

Bothwell, 30 miles southeast of Petrolia in Kent County, was producing some oil. It was in this area in 1856 that Williams and Tripp sank their drill bit, trying to find the mother lode. When the pipe broke in the hole, they became discouraged and went to Enniskillen where Tripp, four years earlier, had started to manufacture asphalt. Charles Oliver Fairbank, grandson of John Henry, has told the story of Bothwell as follows:

Tripp's successors at Bothwell were largely Americans up from the oil fields of Pennsylvania. But their work had scarcely begun to pay dividends when the great flowing wells at Oil Springs opened up, producing oil at a lower price than the 'pumpers' of Bothwell, and the field all but closed down. . . .

Then, one J. M. Lick, late of Pennsylvania, turned in a well producing 1500 barrels before pumping. The rush on Bothwell returned and by 1863 land prices soared. Lick turned up another well pumping 200 barrels a day. During 1865 and 1866 Bothwell grew from a village of 400 to 3,000.

Lick's perseverance, at least, deserves note. No modern wildcatter would even attempt what Lick attempted nearly 90 years ago. With a weight of perhaps 300 pounds. . . he drilled through hundreds of feet of sand, gravel, clay and quick sand. He had more obstacles to overcome than his predecessors had, and won out.

Unfortunately, Lick's luck did not match his

intestinal fortitude. He invested his money in a grand hotel at Bothwell at the height of the boom. But, at the end of the Civil War, oil prices slumped from $12.00 a barrel to $2.00, and the bubble burst, leaving the empty hotel its monument. Years later, the spectacular oil king died a pauper, and to bury him, his friends had to pass the hat.[3]

Although other fields produced oil, none were as involved in bringing order to the industry as was Petrolia. Early wildcatters were independent entrepreneurs who worked alone or with partners producing, refining, and selling their oil in secret, bargaining with consumers for the best price, struggling independently for the best markets. They did not realize that their new-found gold was affected by supply and demand in Canada, supply and demand in the United States, and eventually, supply and demand in the world. Also, imports from south of the border continued to dribble into Canada, and this worried hard oilers and affected the price and the demand for oil.

Fairbank knew, as early as 1862, that producers could not work alone and survive; he had tried to bring them together under the banner of the Canadian Oil Association to establish a decent price for oil, but once the price rose, the oil men went back to working alone and cutting one another's throats in their fierce, competitive style. It was not until eleven years later that Fairbank was successful in uniting the producers under one banner. With the formation of the Home Oil Company in 1873, they

finally stayed together for eight years, realizing at last that there was power in unity.

The refiners, as was shown earlier, were even more difficult to organize. The Refiners' Association in 1868 was a beginning, formed to regulate the selling price of kerosene and to promote foreign trade. In a year, this organization disbanded. As one historian put it, "combinations in business which depended upon a gentleman's agreement, while they often showed signs of success at the outset, frequently failed when individual members secretly undercut the agreement they had signed by selling oil privately at discount."[4]

The London Oil Refining Company that Englehart had formed in the mid-1870s lasted just under two years, but the refiners were then beginning to realize that oil was big business and the little man could not endure unless he had the clout of a big organization behind him.

In the 1870s, Rockefeller's Standard Oil Company used rough, tough, often underhand, tactics to dominate the refining business in the U.S.A. By 1879 Rockefeller controlled ninety percent of refining operations south of the border, and his henchmen had been to Canada to study all facets of the industry in London, Oil Springs, and Petrolia. Englehart and a number of London refiners knew that the time had come to unite under a strong banner in order to keep Rockefeller out of Canada.

In London early in 1880, Englehart called a meeting of a number of refiners with a view to amalgamation. They were well aware of die-hard

The original headquarters for the Imperial Oil Company, Limited in Petrolia, 1882.

British elements in Canada, and they decided that their new company should reflect its British-North American roots; they called it Imperial Oil Company, Limited. Capital was set at $500,000, a hefty sum in those days, and this was divided into 5000 shares of $100 par. Of this, 2,928 shares were issued. Englehart purchased 577; Herman Waterman, 420; and Frederick Ardiel Fitzgerald, 292. Other shareholders were allocated from 10 to 146 shares each.[5] As John Ewing notes in his history of Imperial Oil, "few, if any, of the group came up through the fields by way of the drill or the derrick, and some were more interested in other things than oil at the time the company was founded, contributing mainly capital to the venture. For the majority, however, oil was the primary concern, or perhaps more honestly, the desire of profits through oil."[6]

Founders named Frederick Ardiel Fitzgerald, London businessman and oil refiner, as president of the new company; Englehart as vice-president; and William Melville Spencer, who owned a London still, as secretary. Other original shareholders were: Joseph S. Fallows, William D. Cooper, William English, John Geary, Edward Hodgins, Thomas D. Hodgins, John R. Minhinnick, Thomas H. Smallman, Charles Norman Spencer, William Spencer, John Walker, Herman Waterman, and Isaac Waterman.[7] As John S. Ewing notes, "their backgrounds, birthplaces, racial origins and religions were dissimilar; their numbers included Jews and Gentiles, Americans, Scotsmen, Germans and native-born Canadians; they had been farmers, merchants, lawyers, railway men. In ability, too, there was marked difference, for some were to lose the money which they made from Imperial and die in bankruptcy, while others were to go on to other successful ventures and into public service.... Although they did not realize it at the time, these men had formed what was then, and what was to remain, Canada's largest oil company."[8] Their founding aims were two-fold: "to produce better products (in those days mainly lamp oil, axel grease and other lubricants, wax and candles) and to develop new markets.... Their charter was to find, produce, refine and distribute petroleum and its products throughout Canada."[9]

They decided that the site of their operation would be Englehart's refinery in London. As manager of this operation, Englehart, for the next three years, would be a busy man, for he also operated his Silver Star Refinery in Petrolia. However, at age thirty-three, he was a confident, wealthy businessman who chose capable men to work in key positions to help him run his empire.

The oil men who founded Imperial were quick to see opportunity in a new field, and they were well able to cope with it financially and technically. Historian John S. Ewing has outlined their accomplishments:

Herman Waterman had been a clerk in a London clothing shop when his interest turned to the new field of oil refining. His brother Isaac, who had come from Germany to Canada in 1858, some time later than Herman, was first a partner of William Spencer; then, with his brother, he established the Atlantic Petroleum Works in London in 1866, before the Englehart association. Herman Waterman confined his activities exclusively to refining, but Isaac, who was mayor of the village of London East, was prominent also in other companies, including the London Street Railway Company, City Gas Company, and the London Steam Supply & Manufacturing Company.

The Spencer family had begun refining oil in Woodstock as early as 1862, and William Spencer was reported to have been associated with James Miller Williams. Apart from the brief partnership with Isaac Waterman, he and his sons — William Melville Spencer and Charles Norman Spencer — do not appear to have been otherwise connected previously with the remainder of the Imperial men, nor do they seem to have been interested in producing or in other businesses.

William Cooper, William English, John Geary, the brothers Thomas D. Hodgins and Edward Hodgins, and John R. Minhinnick were purely oil refiners, like the Spencers. John Geary had been a lawyer, and Thomas D. Hodgins a carriage and wagon maker in 1868, but most of the others were already refining by that time. By 1874, they were all active refiners, and English and Minhinnick, in 1880, were connected with a London company called the Victor Oil Works, while Minhinnick was also a partner with John Geary in another concern. Cooper does not seem to have had close business connections with any of

Frederick Ardiel Fitzgerald, first president of Imperial Oil Company, Limited, 1880–98.

the group, but his experience in refining went back to before 1868; in that year he was a partner in the Gore Oil Works, and continued to have connections with that firm for some years. . . .

Of the remaining founders, Thomas Smallman and John Walker had been associated for some time in the Canada Chemical Manufacturing Company, in London, which was the first firm to manufacture sulphuric acid in Canada, and in an earlier oil venture, involving English capital, which had failed. They were partners in 1880 in the Mutual Oil Refining Company, which in 1879 had purchased the Erie Petroleum Works in London.

Smallman maintained his interest in the Canada Chemical Manufacturing Company until it was later absorbed by an American firm, after which he took an active part in the London Street Railway Company. Walker, who had been a vice-president under Sir Hugh Allan of the Canada Pacific Railway Company became actively interested in politics. He sat for some years in the House of Commons at Ottawa, and was finally appointed registrar of Middlesex County.[10]

Fitzgerald, the first president of Imperial, was born 16 October 1840 on a farm in London Township. He was the son of John and Rebecca Fitzgerald, and grandson of Edward and Margaret Fitzgerald who, with seven children, had emigrated from Ireland to Upper Canada in the late 1700s.[11] When Imperial was formed, Fitzgerald was an associate in refining with Joseph Fallows, and both were partners in F. A. Fitzgerald & Company, a lumber firm. "Fitzgerald had begun his business career in London as senior partner of Fitzgerald & Scandrett, wholesalers and retailers of groceries, wines and spirits," John S. Ewing writes, and "he retained his interest in this firm and in the London Furniture Company, which he founded in all probability, as an offshoot of the lumber company."[12]

Bearded, dark-haired, of medium height, Fitzgerald, like Englehart, drove himself relentlessly, working eighteen hours a day. He built the Fitzgerald Block, an impressive building in downtown London, and a few years later, probably 1865, he became interested in oil production and refining and built refining plants in Petrolia and London. He was married and had four children.

Fitzgerald was a Wesleyan Methodist like his father and grandfather, but there is no evidence that he possessed the same religious fervor that Hugh Nixon Shaw had possessed. Like Shaw, he did have a certain quiet, persuasive Irish charm. At a later date when he was in charge of sales for the company, this Irish charm helped to grease the wheels of business relationships and make profits, too.

Englehart also was persuasive in his own, sometimes stubborn way. Both men knew where they were going in life and in business, and they made a good team.

The oil refinery district in Petrolia in the early 1890s.

An Act of God re-directed the course of Imperial Oil in the first three years. Lightning struck the London plant. Resounding explosions, fire, and smoke shook up the young city. After the holocaust burned itself out, there was nothing of the refinery left to salvage. Just prior to this catastrophe, Imperial had asked the London City Council to give $20,000 to assist in building a 60-mile pipeline to Petrolia to help defray the high cost of transportation of crude oil.

There were fifty-two refineries in London at that time, and fires had been an everyday occurrence. Council members were fed up with the smoke and smells created by refining, and they adamantly refused to give any help to build a pipeline. Undaunted, the new company moved its whole enterprise to Petrolia where it continued operations in Englehart's Silver Star plant. Spread out like an octopus on 50 acres of land, and needing five hundred men to keep it operating at full capacity, this refinery was really the launching pad for Imperial Oil Company, Limited. And little Petrolia, the town that was not meant to be a town, but just a passing dream in a man's life, was well on its way to becoming the oil capital of Canada.

THE OIL EXCHANGE

There is no evidence that Jake Englehart approached John Henry Fairbank to ask him to be a member of Imperial. Certainly, Fairbank would not have wanted to be included in this refining operation. Although he had long urged producers to amalgamate for strength and power, he did not like cumbersome corporations where the little guy and the aims could be lost in a tangle of paper work. He was very much a self-made man and an independent thinker. Also, like all producers, he did not like refiners. Producers blamed refiners for undercutting the price of crude oil. Refiners, in turn, blamed producers for not restricting production to create demand so that the refiners could ask big prices for the refined product. "Of course the truth was that the petroleum industry, like any other, could neither stay still nor go forward in a vacuum," John S. Ewing writes in his history of Imperial Oil. "Over and above the bickering of the two rival factions, larger forces were at work beyond their control. The little petty capitalists of Western Ontario were simply pawns in a bigger game."[1]

Fairbank was well aware of the complex workings of the oil business, though. Nearly twenty years had elapsed since he had come, a penniless immigrant, to Oil Springs where he had fallen in love with oil. In that time he had learned all there was to know about petroleum, and had, through careful transactions, turned himself into a wealthy man. No doubt he thought that Imperial would, like all combinations before it, go belly up within a few years. He and Englehart did not always agree with one another on the manner of conducting petroleum business. In fact, they did not go into business together except for one venture, the Crown Loan and Savings Company which was incorporated 30 January 1882 with Fairbank as president and Englehart as vice-president. After Fairbank's death in 1914, Englehart became president of this banking concern.[2]

Although Fairbank did not like refining, he was connected with at least seven different refining projects during his lifetime. His most important refining operation was the Home Oil Company, formed in 1873 with fifteen shareholders. "The refinery was built at a cost of $37,000," Edward Phelps reports, "and could process three thousand barrels of crude oil per week, which represented about one-half of the current production of crude and nearly all of the current demand for refined oil in the country."[3] Home Oil was a thorn in the side of London refiners who, because of its dictates, often had to beg for crude and pay the price asked. But by 1881 Imperial, still located in London, was powerful enough to purchase Home Oil. No doubt the price was right or Fairbank would not have sold. Also, a political career began to loom large on the horizon. Perhaps he thought he should let the refinery go in order to conserve his strength for a new undertaking.

During the early 1870s he had become a British citizen.[4] As Alexander Mackenzie's good friend, he had always championed the Liberal cause, and in the spring of 1882 he was unanimously chosen Liberal candidate for East Lambton. On election day that year he slid into office with a resounding majority.

This election victory came only a few short months after a family tragedy; his eldest son, Henry, committed suicide at the University of Michigan.[5] Henry had graduated from the University of Toronto in 1880, and then enrolled in the Medical School of the University of Michigan at Ann Arbor. Charles, his younger brother, was a commissioned officer in the Royal Artillery, training in Woolwich, England for a career in the British Army. When he received word of his brother's death, he returned home immediately. On a cold, snowy day in February the funeral cortege wound slowly through the streets of Petrolia from Fairbank's home to the Anglican church and from there to the family plot in Hillsdale Cemetery. The local paper reported that "it was the largest funeral ever seen here. Over 150 vehicles were in the procession, besides a large concourse of citizens on foot." Englehart and VanTuyl were honorary pall-bearers.[6]

Henry's death was a tragic blow for Edna who never seemed robust at the best of times. In a letter to Charles six years earlier when he was at school, she had written: "I had a bad spell last Friday for a little while. My heart doesn't seem to work right sometimes, poor old heart (about 500 years old if suffering & sorrow can give age)."[7]

To give moral support to the family during this crises, Fairbank urged Charles to stay home following the funeral of his brother. He remained in Petrolia for the next seven years. In this way he was able to help his mother over the difficult mourning period, and was also able to keep an eye on the many family businesses while his father was in Ottawa. The two men became close friends during these years when the demands of government weighed heavily on the elder Fairbank's shoulders.

Politicians are not as well liked in other parts of the country as they are at home. This is often a difficult pill to swallow for the person who is new at the job, and Fairbank, now fifty-one years old, learned the lesson early in his career. In September 1884, he traveled with Alexander Mackenzie and other prominent Liberals to Winnipeg. In a letter to the editor of a Winnipeg paper, Fairbank's speeches and mannerisms were portrayed derisively:

Sir — In his speech last night at the Mackenzie reception, Mr. Fairbanks, M.P. for some county or part of a county in Ontario, accused the press here of persistently calling him a joker. I think the press must have been indulging in gratuitous hyperbole to have bestowed upon him so dignified an appellation. It is true he apes the role; it is also true that he is possessed of all the attributes of a clown, except that of being "funny," but as that happens to be the chief qualification in the calling of a buffoon, its want is a somewhat serious delicacy. But Mr. Fairbanks need not despair. We have a local professor of the art; and I have no doubt that for a consideration Richard Burden, Esq., will give such instructions as will develop the Ontario mountebank into a really accomplished buffoon.[8]

Fairbank lost the election three years later to George Moncrieff, a Conservative. But no doubt he was happy when all the votes were counted and he was out of government, for then he could get back to the oil business.

While still a member of parliament, Fairbank helped to form the Petrolia Oil Exchange, so that buyers and sellers could meet at the Little Red Bank at noon each day to discuss business. As recorded in the Board of Management minutes of the Exchange for 1884-85, "Thirty-four producers and refiners signed up as charter members before the Exchange opened for business on Dec. 23, 1884; many others subsequently joined, including twenty-two in 1885."[9] The Exchange controlled "through inspection the quality of oil offered for sale under its auspices. . .The officers enforced stringent rules to ensure that all transactions were made in good faith, and to prevent anyone from manipulating the price of crude for private advantage. Warehouse receipts from the various tanking companies usually changed hands during transactions."[10]

In good weather, the bulls and bears of the petroleum industry met outside near the picket fence at the front entrance to the bank. Later, an elaborate Oil Exchange building was erected on the main street to accommodate this very important group of oil men. The Petrolia Oil Exchange was one of the town's most famous institutions. It kept its finger on the pulse of the oil industry at all times, and it endured for thirteen years. But as Edward Phelps observes, "if oil men had hoped that centralizing the sale of oil through the Exchange would raise the price appreciably, they were doomed to disappointment. On its opening day the Exchange saw the price of oil set at the low of 75 ¼ cents a barrel. In succeeding weeks the price rose and fell only slightly, in response to normal demand. The Exchange, however, may well have prevented complete demoralization of the market."[11]

Probably one of the most important aspects of the Exchange was that at long last a combination of producers *and* refiners were talking and doing business *together*. Within the ranks however, there was some bickering that occasionally rose to open warfare, as happened on a wintery day in January 1892. A letter to the editor of the *Petrolia Advertiser* gave readers a blow by blow description of the fracas. The "players" mentioned — Simmons and Cameron — were oil producers who were also involved in municipal and county politics:

Dear Sir: —I herewith hand you a report of a scene of comic opera.

Scene: — Oil Exchange Office.

Time: — About noon on Tuesday, 29th.

Enter members, to the number of about twenty and take seats around the room. 12:30 bell rings. Curtain rises.

Mr. D. Cameron — Passing swiftly across the room, stops suddenly in front of Mr. Simmons, gauge of face shows 100 lbs. pressure, delivers himself thus with loud voice and lungs expanded:

"You are a sneak, a coward, a liar. I am a superior man to you, both physically, intellectually, morally, on the platform, in the press, and — you — I can lick you, I can mop the floor with you."

Mr. Simmons — With aggravating coolness, "Truly you are a most wonderful man — "

Cameron, with gauge indicating 125, "You're a liar, your family are liars. There is not anything too mean or contemptible for you to do. You lied on the platform about me when I could not answer you. I can lick you. I am a Cameron and know no fear — "

Mr. Simmons — with great courtesy. "Yes, you were kind enough to mention it before."

Mr. Cameron — Gauge showing 150 and getting red in the face. "Yes and I mean it. You can try it any time you like. I can lick you quicker than fire would scorch a feather, you mean, contemptible hound. Come out here and I will do you up in two rounds."

Mr. Simmons — with hands in his overcoat pockets and a quiet smile upon his face. "Do, my dear Mr. Cameron, try and control yourself; you are getting quite emotional. I fear if you continue in this way you will get excited."

Mr. Cameron — Gauge 175, face redder. "Mr. Simmons, I again tell you, you are a liar, a sneak and a coward. Your actions prove it. Your family are sneaks and liars and are hated by everyone, whilst my family's society is sought after by the best families in town; you are craven and brutal and cannot conduct yourself with respect in the company of respectable people."

Mr. Simmons — "My dear Mr. Cameron, let me advise you to be careful, you really look as though you had a rush of blood to the head, there is great danger of bursting a blood vessel in the place where the brain should be."

Mr. Cameron — At the 200 mark, laying down his umbrella and pulling up the sleeve of his coat, "Mr. Simmons, if you ever mention my name again from the public platform, or speak of me again at any time, you cowardly, miserable, stinking liar, I will lick you on sight. I'll trample you in the mud. I'll skin you alive. I'll draw and quarter you; don't speak to me again. I am your superior in every respect. I can lick you, and be you Warden or Mayor, I'll do it. Physically, mentally, morally, on the platform, in the press, or prize ring, you are no match for me. I am a Cameron and know no fear."

Mr. Simmons — That aggravating calmness still to the front. "My dear Mr. Cameron, you're very excited state greatly alarms me. You must, my dear friend, try and cool yourself off a little, do be advised. Get ice to your head. I am sure it is much over-heated. You look really as though you were convulsed; do try, my dear friend, and reconcile yourself. I know you promised to vote against me, and I am sure I won't feel bad about it if you will only try to take care of your very precious self. Lives like yours are not so plentiful that you should take so much liberty with them, besides you might get hurt sometime when one of those great outbursts of wisdom has to find vent.[12]"

The writer titled his letter "Almost a Tragedy" and signed it "Othello."

WILLIAM MCGARVEY IN GALICIA

There is no doubt that Jake Englehart would like to have included William McGarvey in Imperial, but McGarvey's involvement with oil was taking him away from Petrolia. In 1875 he was a member of a federal survey party that went to Northwest Canada to explore for oil. Professor John Macoun of the Geological Survey of Canada went into the Athabasca region of Northern Alberta (then the Northwest Territories), and this was, no doubt, the expedition that included McGarvey. There is no evidence that they found oil, but they did notice tar springs on a trip up the Athabasca River.[1]

McGarvey owned oil lands in Enniskillen with his father, Edward, and in 1879, the Petrolia paper advertised the sale of these lands.[2] It would seem that McGarvey was beginning to loosen home ties in anticipation of a career elsewhere. Yet that year he was elected warden for Lambton County and for twelve months was involved in county politics. It would be another year before he went wildcatting in Europe.

In 1880, John Simeon Bergheim, a British engineer who was interested in oil exploration, traveled to the oil regions of Pennsylvania to find a crew of well-trained men who might be interested in wildcatting in Germany. He had little luck in finding a crew in the U.S.A. Drake's find in 1859 pointed the way for oil exploration on a grand scale in America. In 1871 the Bradford, Pennsylvania-New York field was found, and this sent prospectors scurrying in that direction. Towards the end of the 1870s there were drilling activities in Ohio, and speculation ran high that there was oil all the way from Virginia to California. American oil men were busy developing new fields at home, and foreign fields across an ocean did not lure them the way they lured Canadians. Canada was a young country, scarcely thirteen years old, and she had other priorities more important than oil, like building a railroad coast to coast. Exploration for minerals would have to wait. The great gas field at Turner Valley, Alberta would not blow in for another thirty-four years, and the first big wild well at Leduc

was still sixty-seven years away. The only oil wells commercially active until 1914 were those in Enniskillen, and from 1874 onward a recession in Canada forced drillers to find work across an ocean. Pennsylvania oil men were tuned into this shift of interests in Canada, and they told Bergheim to look for drillers in Petrolia.

In 1880 McGarvey was restless. His government survey job had introduced him to adventures beyond Enniskillen and the challenges that these adventures offered. He and Bergheim liked one another at their first meeting in Petrolia. They were both self-made men and independently wealthy.

In order to drill in Germany, the men had to join the Continental Oil Company, and McGarvey was soon named as the company's director. He took a crew of hard oilers with him from Petrolia, and they set up their Canadian Pole Drilling Rig near Hanover, Germany. They did not find oil at this location, so they decided to move to Galicia, then part of the Austrian Empire. McGarvey had studied the European-Asian oil situation and he knew that oil was mined in Galicia in the 18th century. There were hand-dug wells in Sloboda, Boryslaw, Drohobycz, and throughout the foothills of the Carpathian Mountains. As one German historian recounts:

In the beginning of the 19th century, Galicia extracted 6900 litres of oil yearly. In 1810, at Boryslaw and Drohobycz, the first attempts were made to distill crude oil in order to refine it into lighting oil. Up to that time, such an attempt had

Illustration of a drilling rig from an Oil Well Supply Company catalogue.

not yet been made anywhere in the world. The entrepreneur was the tax official J. Mitis, who was so successful that in 1817 the city council of Prague decided to light the streets of the city with the new lighting oil, known at the time as 'naphtha'. The city signed a contract with the refiners to supply 160 hundredweights of naphtha at 34 guilders per hundredweight. However, it later canceled the plan for fear of fire. Thus the further refining of naphtha did not occur, though the demand for oil increased steadily. . . . In the year 1853, a certain Abraham Schreiner succeeded in making a completely clear and transparent distillate. Together with the pharmaceutical chemist Lukasiewicz, he set up the first real refinery to produce this product. The first lighting with this lamp oil, in the hospital of Lemberg, was such a success that soon after a whole series of small refineries came into being in Klisch, Gorlice and Polanka. . . .

Prior to 1862, the oil was extracted simply from manually operated wells, with a diameter of one metre square. The oil was found about 100 metres deep, and for protection against gas explosions the shafts were covered with planks. In 1862 a start was made at Boryslaw and Bobrka with hand-drilling, and the first mechanical drilling took place in 1867 with the cable drill, popular in Pennsylvania at that time. The first attempt with cable drilling [the spring-pole method] was made at Klentschany, but the rock there proved to be unfavorable to this method. Better results were obtained with the steam drill, which was introduced at

Mentschina, slightly later, though the faulty tubing necessitated them to stop drilling at 200 metres.

In the year 1882, McGarvey introduced the so-called "Canadian Drilling Method" in Galicia. This method proved to be extremely useful for the geological conditions, and with it the deepest levels could be reached. Because of this the production rose, especially in Boryslaw which became the center of the Canadian drilling method, and where much capital was invested.[3]

Petrolia historians Charlie Whipp and Ed. Phelps note that "the skills the Petrolians used in foreign fields were those they had learned at home. Frequently they employed black ash rods rather than cable when drilling in unknown formations. The rods kept the drill bit falling in a straight line. When they had rarely gone deeper than 400 feet at Petrolia, the foreign drillers employed their pole tools to go down several thousand feet in Australia (and Austria). Their drills and other pieces of equipment were designed for them by the Oil Well Supply Company and every foreign driller knew whom to write to when he had technical problems."[4]

As was mentioned earlier, McGarvey made his own renovations to the original Canadian pole tool method of drilling. He also made the rig portable. The Oil Well Supply Company began to make drill bits and other oil tools in 1866, and it is still doing a booming business today in the 1990s. The company became so famous that, in 1880, the two original partners opened another outlet in Stockport, England.

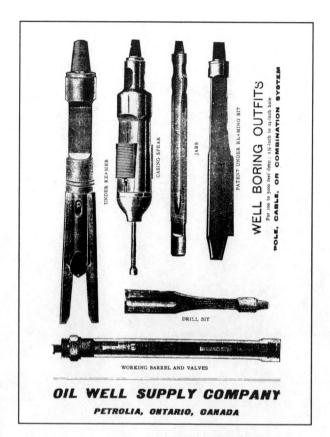

Another catalogue illustration of equipment available from the Oil Well Supply Company.

*Laborers in the Galician oil fields
of Austria in the 1870s.*

McGarvey was equipped with a crew of well-trained drillers, and he knew where to get supplies as quickly as possible. He was certainly the right man in the right place when he and Bergheim arrived in the heart of the old oil district in Galicia — Boryslaw and Tustanowice. They must have been surprised when they first set eyes on this beautiful Carpathian Mountain district, described as follows by German historian Martin Pollack:

They were looking at a cratered landscape. The oil pits, often at a distance of just a couple of meters from each other, had narrow shafts dug straight into the ground to 150 meters deep. In these the crude collected at the bottom, and was scooped up with a bucket as from a well, a bucket which was let down with a rope running over a wooden windlass. The most important raw material of this type of mining, apart from the oil itself, came from the willows which grew in the thousands along the Trudnica, the Bar and the Tysmienica Rivers. The shafts, cut into sandstone, had walls covered with wicker work; the winding mechanism was formed by two forked willow trunks driven into the ground next to the shaft; a third, thinner branch, was put on top of the forks by way of reelshaft. The winding rope, too, was woven out of willow twigs, as well as the handle of the wooden drawing pail.[5]

McGarvey had never seen oil fields as primitive as these. Even when he arrived in the early 1860s in

Oil Springs, wells were drilled by the spring-pole method and pumped by horse or steam power.

He began to do exploratory drilling, and within six months his crews touched off a gusher that spewed out 30,000 barrels of crude a day. Bergheim and McGarvey were as excited as children with a new toy. Working with drillers who were well-trained in caring for wild wells, it still took them four days to control the well. They built makeshift storage tanks and a still, and quickly ordered further supplies from the Oil Well Supply Company's new depot in England. This first wild well assured their success in Galicia.

In the next nine years they drilled 370 wells with a combined depth of 100,000 meters. The advantage of using the Canadian Drilling Method was that they were able to reach a considerable depth surely and quickly. They could drill down to 24 meters in 24 hours. The Bergheim and McGarvey enterprise was considered the most important drilling and refining company in Galicia from 1883 to 1914.[6]

Independent producers flocked into Galicia following their successes, but McGarvey and Bergheim bought out many of these small companies. A few processing outfits moved in, and, as had happened earlier in Oil Springs and Petrolia, competition became fierce, but McGarvey was well-tuned to this cut-throat business. He was not a reckless entrepreneur, but careful, well aware of risks.

Twenty-five hundred miles away, on the shore of the Caspian Sea, the Russian city of Baku, well-known for its oil and gases in the days before Christ,

Oil derricks erected on the foothills of the Carpathian Mountains in the 1870s, the site of William McGarvey's foreign enterprise.

was suddenly coming to life. As early as 1873, Robert Nobel, brother of Alfred Nobel who discovered dynamite, bought an old refinery in Baku and began to work it. In 1876, his brother Ludwig joined him, and together they became the most efficient and successful refiners in the area. In the early 1880s, when McGarvey was having spectacular successes in Galicia, there were two hundred refiners in Baku, and the Nobel brothers dominated the scene.

Baku was a city more influenced by eastern culture than western. It had once belonged to Persia, and an old quarter within its limits reflected this, with narrow streets, mosques, and minarets. In this ancient city, the Nobel brothers were cut off from getting their oil out to markets because railroads to the Black Sea 1000 miles away were non-existent. In 1880, Russia was able to solicit the financial help of Baron Alphonse de Rothschild, a banker in Paris, to build a railroad; three years later, this railway to the port of Batumi on the Black Sea was in operation.

McGarvey and Bergheim were tuned into the news of this competition in Baku, but they held a giant position in the Austro-Hungarian Empire in oil production, refining, and marketing. They were determined to stay on top.

McGarvey was strictly a family man. He, himself, had grown up in a large, closely-knit family, and no doubt he did not like his bachelor living experiences in Galicia. About 1884, when he knew that his operations would be a success, he sent for his wife, Helena and their children, Nellie, Frederick, and May. The family set up housekeeping at first in the quaint Galician village of Borislaw where the beautiful Carpathian Mountains offered a backdrop. In the summer, these mountains rang with the haunting music of shepherds who played the flute while they guarded their flocks of sheep. In the winter, the temperature often dipped to 30 and 40 degrees F. below zero, but it was a time for peasants and villagers to dance and sing. However, the people were poor, working small farms that belonged to landlords who lived miles away, in big cities.

At this time, Franz Joseph, the last important ruler of the Habsburg dynasty, was Emperor of Austria and King of Hungary. At a later date, Franz Joseph's only son and heir, Archduke Rudolf, committed suicide; nine years later his wife, the Empress Elizabeth, was assassinated by an Italian revolutionary; and in 1914, his nephew, Francis Ferdinand, heir to the throne, was assassinated by a Serbian nationalist, touching off the First World War. These were troubled times in the Austro-Hungarian Empire. At a later date such tragedies would have a mighty impact on the life of William H. McGarvey.

IMPERIAL SUCCESS

While William McGarvey built a petroleum empire in Europe, Jake Englehart and Frederick Fitzgerald were busy building a refining empire at home. In order to honor the Imperial Oil charter "to find, produce, refine and distribute petroleum and its products throughout Canada," they decided to look towards the Canadian west rather than the competitive east. As the authors of *The Story of Imperial Oil* notes, "even after the last spike of the Canadian Pacific railway was driven in 1885, the company's products still creaked into remote settlements — mining and lumber camps, Hudson's Bay Company posts — in oak barrels aboard Red River carts, wooden wagons originally used for buffalo hunting." Imperial offered homesteaders the grand sum of $1.25 on each empty oak barrel when the barrel was returned to the company, but few were returned. They were more valuable as washtubs, rain barrels, and, suitably modified, armchairs. "As a marketing strategy, Imperial's scheme to supply the west was brilliantly successful," the authors of *The Story of Imperial Oil* write. "By the end of its first decade the company had a chain of supply depots stretching from the Great Lakes to Vancouver."[1]

One recruit who began to work as a shipping clerk at the Petrolia plant said that Fitzgerald and Englehart were kind men, but paternal in their attitude towards workers, as John S. Ewing recounts:

Fitzgerald had a beard that was extremely long and luxuriant, even by comparison with other beards in this hirsute age. . . . When he [the new employee] reported for work, he was taken in to speak to Englehart. . . who adjured him to work hard and to be all eyes and ears inside the office so that he might learn as much, as quickly as possible. Outside of it, however, he was to know nothing of the business; such advice reflected the deeply felt conviction that Imperial's affairs concerned Imperial only. The company was not unique in that respect; secrecy was a universal goal of private business in the nineteenth century as the petty capitalist training of most businessmen died hard. Imperial, like other companies, failed to recognize that it became vested with some public interest as soon as it became the dominant manufacturer and marketer of petroleum products in Canada.[2]

Englehart, apart from duties as vice-president, was general manager of the refinery, and Fitzgerald, besides being president, was in charge of sales and business in the branches. As Ewing explains, "for many years Imperial had contracts with the Dominion government for the supply of lighthouse oil, and with the major railway companies for their requirements of refined and lubricating oils; these requirements were filled directly from Petrolia, and responsibility for negotiation of the contracts was carried on by Fitzgerald."[3] The lighthouses that were serviced lay off the coast of Newfoundland, and in the area of St. Pierre and Miquelon, rugged islands in the same stretch of water.

Fitzgerald could be strict when dealing with the branches. In 1892 he reprimanded H. A. Drury, manager of the Saint John branch. A customer had bought $3,000 worth of merchandise, then had run off in an attempt to avoid payment. Drury had chased him, eventually catching the culprit and, over several months, extracting payment in full. Fitzgerald wrote:

You are not at liberty to open large accounts and extend any long or large lines of credits to anyone without first consulting this office. . . . I desire to have it distinctly understood as to large lines of credit and when an a/c is running larger than it should, and being slow of pay, we want to be immediately consulted, and while we are just as liable to make a mistake as you are, still we want to have a hand in it and then you will thus clear your skirts.

In the next paragraph, Fitzgerald slightly modified his strict tone:

I am sorry to have to write in this spirit, but one feels better after saying just what one means, and I trust you will take it in the spirit that it is given. Your associations with the Co. for so many years have been to a great extent highly satisfactory, hence you might think such a severe reprimand was uncalled for, but — it is better expressed than harbored and the point as to extending lines of credit is one that I want borne in mind and carefully carried out. [4]

Fitzgerald employed traveling salesmen called "drummers" to help him with the burden of sales, John S. Ewing reports. "Travel throughout the field was generally by horse and buggy or cutter, to those towns which could not be reached by rail. Even when a town was served by train, trips meant dirt and smoke, spasmodic connections, long waits in dreary depots, and all the doubtful pleasures of a small town hostelry in the 1890s. In the extremes of climate common to most parts of Canada, life on the road was not for the weak, but it must have had its compensations; there are no recorded difficulties in obtaining salesmen, and the turnover among them was low."[5]

Englehart operated his refinery with a strict code of ethics. He insisted that all parts of it remain neat and clean. Safety was a priority. There had been so many explosions in his London plants, resulting in the death of one worker, that he had learned the hard way that safe operations paid off. He initiated workmen's compensation claims long before governments considered doing it.[6]

As a result, the authors of *The Story of Imperial Oil* comment, "the Petrolia refinery became a model for its time. . . . Imperial pumped oil from its own wells, piped it to the refinery or hauled it by horse-drawn wagon and turned it into kerosene, lubricating oils, axle greases, waxes and candles, shipping the products in barrels made in its own cooperage with wood from its own woodlots. It even made its own square oil tins, complete with faucets and screw caps, which customers used all over Canada and overseas. The company grew at such a rate that by 1893 there were 23 branch offices from Halifax to Victoria."[7]

Employees at the refinery and branch offices were dedicated to their jobs and dedicated to Imperial. There is a macabre story told about one employee who spent fifteen years working in the Petrolia wax plant. He left instructions that after his death "his body be encased in the company's paraffin wax, and unearthed 50 years later to demonstrate the preservative powers of the product." His request was partly fulfilled, but as one observer noted, "his body kept floating upward in the coffin and was held down by wooden sticks until the wax cooled. But the tip of the nose came through the surface. They floated another covering of wax over the surface to cover the exposed nose after it cooled."[8] The wax-covered body was never disinterred and presumably rests today in Petrolia's Hillsdale Cemetery.

Although workers were dedicated to Imperial, so was management. As early as 1884, the founders decided to deal with the exasperating skunk smell of the oil. They hired Herman Frasch, one of the original petroleum chemists who was also involved in oil refining. Frasch came to the United States from Germany in 1868; he studied oil refining in Pennsylvania and opened his own research laboratory in 1874

Imperial Oil Company refinery stills at Petrolia in the 1890s.

in Philadelphia. Robert Page in his history of the Canadian oil industry explains that "Frasch was made not only an employee of Imperial, but a shareholder and director in the company. His salary was equal to that of the company president. Imperial obviously viewed his work as critical to their future. Unfortunately, Frasch left Canada before his work was completed and patented; the right to his important inventions were gathered in by Solar Refinery, a subsidiary of Standard Oil. There is only conjecture on the reasons for his departure, but Imperial officials remained bitter for years afterward."[9]

In 1885—about the time that Frasch left Imperial — the Lima, Ohio oil field became active, and the oil had a sulphur content like the Ontario oil. Standard had interests in the Ohio field, and therefore a man like Frasch would be worth his weight in diamonds.

The sulphur in Ontario's kerosene remained a problem until thirteen years later when there were unforeseen changes in Imperial Oil Company, Limited. Then the Frasch invention was used to rid Enniskillen oil of its skunky smell. In her study of frontier mining technology, Dianne Newell explains that this method "was based on the principle of a reaction between metallic oxide and the sulphur contained in crude mineral oils, but Frasch went one step further and added copper oxide to the already saturated solution of crude oil and metallic oxide. Not only was the illuminating oil rendered sweet, but also the copper oxide could be recovered for reuse."[10]

Imperial's early years were successful. In the first six months, Robert Page reports, the company "achieved a profit of 116,000 dollars and paid a dividend of 19 dollars per share (with par value of 100 dollars); in 1881 the dividend was 18 dollars per share; and 28 dollars in 1882."[11] But there were troubled times ahead for directors and shareholders.

BARONIAL BUILDINGS

Jake Englehart was the most eligible bachelor in Petrolia in the 1880s. During that decade he turned forty, but he was still youthful in appearance. Dave Stothers, a Petrolia tailor, said that Englehart was always "immaculately dressed. He preferred dark suits and wore thick-soled shoes with a high shine, a high starched collar, eyeglasses with a broad, black ribbon; he had a moustache, a Vandyke beard, and invariably sported a flower in his buttonhole. He was a nice man, but all business — all business, no fooling. [He] used to bach here one time; him and Ed. Kerby, the first mayor, bached together in a frame house."[1]

In the summer of 1887, Englehart's father died. An obituary in a New York city newspaper read: "Joel Englehart, who died yesterday, was the father-in-law of ex-Judge Abram J. Dittenhoefer, the well-known lawyer of this city. Mr. Englehart was in the 81st year of his age, and had been living for some time past at Laurelton Hall, Cold Springs Harbor, Long Island, where he died. He has two sons, Jacob L. and I. Albert Englehart; the latter was, for a long time, associated with Judge Dittenhoefer. The deceased was well-known on Long Island and in Ohio where he had, when young, transacted considerable business."[2]

Englehart went to New York to attend his father's funeral. Rumors had persisted for a long time that, when in New York, he often dated an attractive young Jewess whose father was a wealthy lawyer. This could have happened, for Englehart had money, prestige, and power, and no doubt women were attracted to him. But shortly after his father's funeral, he began to be seen around Petrolia with Charlotte Eleanor Thompson, George Moncrieff's young sister-in-law.

Charlotte had grown into an attractive, tall young woman who was active in civic and church affairs. She was well-known as president of the town's Relief Fund which gave help to needy families. Although sixteen years younger than Englehart, she had always seemed older than her years, perhaps because she was the oldest of the children in the Moncrieff family. She never officially changed her name, but the 1881 census listed her as Charlotte Eleanor ("Minnie") Moncrieff and not Charlotte Eleanor Thompson. That year she was also listed as seventeen years of age, and living with George Moncrieff and his wife, Isabella, and their five children.[3]

Portrait of Charlotte Eleanor Thompson.

Victoria Hall in the early 1890s.

Petrolia, in the 1880s, took on a more permanent look, as described by civic historians Charles Whipp and Edward Phelps:

Brick buildings rose, where before there had been mostly wood. Many went to Port Huron occasionally and the merchants of that city advertised heavily in the Petrolia papers. . . . Virtually every form of retailer had a store in Petrolia. Dempsey's employed a dozen clerks, Lancey's Folly was an even larger dry goods store. The town's two photographers did a land office business selling photographs of local buildings and special occasions. . . . The VanTuyl and Fairbank hardware store was one of the largest of its kind in Ontario, doing an annual business of $250,000. Traffic on the main street was constant and heavy. There were many wealthy people here and it was not uncommon for Englehart and Noble to be seen dining in Delmonico's in New York or Corey entertaining visiting artists and performers from Toronto and Montreal.[4]

The main street was paved with wood blocks, and in 1889 the town hall, an elaborate Victorian building of brick, was finished. Called Victoria Hall for Queen Victoria, it housed, on the main floor, the town's council chambers, the jail, the fire hall; and above these quarters, a splendid theater with stage, balcony, and box seats edged with wrought-iron railings.

This theater had been just a dream a short three years before. At that time, Canada's Prime Minister Sir John A. Macdonald visited Petrolia. A reception was held in the curling rink where people were packed in as tight as olives in a jar. The Oil Exchange Hall had been built a decade earlier to accommodate the bickering, trading, and price-fixing in the oil industry, but this hall was not much bigger than the curling rink. People decided that the time was ripe to raise the necessary $35,000 to build Victoria Hall. It was certainly money well spent; the hall was used for political meetings, annual balls, and as a community center by both professional and amateur artists. One thousand wicker seats could be set up quickly for a stage performance, and they could be removed just as quickly to make room for dances, like the annual assemblies that lured the rich and famous.

In those halcyon days of the early 1890s, a professional group of actors from Detroit staged *Dr. Jekyll and Mr. Hyde*, and the famous Canadian native poet Pauline Johnson read her poetry. It is rumored that Shakespeare came to Victoria Hall, when Sarah Bernhardt, the world-renowned French actress, played Lady Macbeth.

About this time, Englehart decided to build a permanent home in Petrolia, and Fairbank planned to erect an elaborate mansion. They were not to be built side by side, these two castles. The Fairbank home would be on the crest of the hill west of the creek on the main road leading into town. Englehart chose a quiet meadow southwest of this, also near the creek. Neither home was in Crescent Park, and jealous rumblings rose from this end of town. Both men were condemned for putting on airs. In fact, Fairbank was heralded as wanting to be King of

Victoria Hall as restored in the 1990s.

*Sunnyside, the castle-like home of
John Henry and Edna Fairbank the
1890s.*

John Henry Fairbank sitting beside the elaborate stairway in his new home.

Petrolia and Edna as Queen. The truth was, of course, that Fairbank never wanted an elaborate castle-like home. He was happy with a cozy frame house or a log cabin. Rumors persist, even today, that Edna urged him to build a mammoth castle for her. Perhaps she wanted to send a message to the Crescent Park end of town that she did not have to live there in order to be a very important person. She was also ever mindful of her Loyalist ties in Niagara, and perhaps she wanted to put on a show for them.

Englehart and Fairbank ignored the envious rumblings. They were building these homes for different reasons, and they were happy with their choices. Fairbank, no doubt, wanted to please Edna. Englehart had plans to present the home as a wedding gift to a future bride.

The bricks for both houses came from a supplier in Ohio, each brick wrapped in wax paper before shipment. All timbers and hardwood were cut from local woodlots and dried for a year prior to construction. The architectural designs were Victorian — often called Eclectic Revivalism, adapting Gothic and Tudor revival styles to the modern Victorian age.

In 1890, May Fairbank, then twenty-one years of age, wrote to her brother who was studying medicine in New York city:

I have a little picture gallery in my mind . . . and have a certain picture or perhaps more than one of each of my friends and loved ones. Now I see Papa in his old — shall I say overcoat? — black hat, with his shoes muddy, trousers turned up and head slightly bent forward walking around the new house. You will see great changes in the new house when you get home, but Papa says we will not be in it until June or July.[5]

Seventy years later when a grandson — another Charles O. Fairbank — was forced to sell the house because the cost of upkeep was too prohibitive, a reporter from London visited the castle and wrote:

Iron gates off Petrolia's main street admit the visitor's car up the long driveway to the main house. As the green and yellow cover of trees and vines part, the great mass of sandstone and brick assumes it shape. Slate roofs tower over 50 feet above the street. The meticulously laid brickwork curves in turrets and wings and the plate glass windows curve faithfully with them. An ancient wisteria vine clambers over the broad front veranda, knotting itself to the wrought iron railings.

When the oaken door swings back, one is presented with a view of the center hall, 60 feet long and 12 feet wide with wainscotting of rubbed sycamore, beautiful as the day it was cut from the Fairbank farms in Brooke Township. To the left is the music room with its grand piano . . . and beyond, the great staircase with its hand-carved banisters moves up to a vast landing where a picture window 12 feet square sends a flood of light over the gleaming sycamore. Five people could walk abreast up this staircase to the third story ballroom where famous orchestras once played to Petrolia's oil rich society. . . .

For quieter entertaining, J. H. Fairbank provided a great living room 32 feet by 27 feet and beyond, a dining room which easily accommodates its 10-foot table. Across the center hall is the library, den and behind these the maid's sitting room. Behind the dining room are the kitchens and a large, lavishly cupboarded room known as the servery.

Fairbank house employed two maids, a cook, laundress, gardener and groomsman — the latter living in apartments over the stable. He and his horses have long since left, and where the carriages were once kept there now reposes a Packard convertible of approximately 1925 vintage. . . . Beside it sits the inevitable Model T. Ford.

Fairbank House has eight bedrooms with adjoining dressing rooms and bathrooms. The walk-in closets have sliding doors of great weight, but which slide easily. There is none of the dimness and darkness in these rooms, characteristic of so many early houses. Some of the windows go from floor to ceiling and in the bedroom wings there are two and three windows in a row, providing excellent ventilation. Everywhere, the woodwork of oak, ash, birch, elm and maple is like new.

The attics of the third floor would, together, accommodate the average bungalow of today. They are filled with antique furnishings, occasionally protected by dust sheets, and with books, letters, documents, broken or discarded clocks — the accumulations of three generations of a great house and its families.

In the basement one wanders through the wine cellar, the fruit cellar, the kitchen cellar, the laundry and heating rooms, the coal cellar and the wood cellar. Two furnaces — one coal and the other gas — heat the house.[6]

The Englehart home was of similar design. The front double doors were oak-paneled, inlaid with stained glass windows. A living room on the main floor featured a cast-iron fireplace of bamboo design with side flues and porcelain tiled hearth. In between the flues there was a stained glass window with raised glass pieces called jewels. The walls of the room were lined with a paneled wainscotting. The ceiling was paneled, and small finials hung from each jointed beam. Below the molding that circled the room there was a frieze in floral design. Both homes were built on several acres of land, beautifully landscaped.

In those days, large estates were duly christened with names. Fairbank and Edna called their new home "Sunnyside." Nobody knows why this particular name was chosen. No doubt Edna hoped that her castle would bring sunnier, happier days than she had experienced in the past. Englehart, probably because his house was built on the crest of a hill overlooking a glen, called his dream home "Glenview."

Many people were in awe as they watched these grand, artistic palaces rise in their midst. One local newspaper commented: "The new residence of J. H. Fairbank is being pushed ahead rapidly and when finished we doubt if there will be anything

*The Main Street of Petrolia
in the 1890s.*

west of Toronto to equal it in excellency of material and architectural design."[7]

However, a few of the nouveau riche called both homes "ugly."One old timer commented that "they're terrible monstrosities, those two barns — and the ballroom in the Fairbank house is stifling hot if you get more than six people in it. The only nice thing about the house is its north tower — if you like towers."[8]

Englehart and his castle Glenview were not as maligned as was Fairbank and Sunnyside, probably because Englehart had the power to hire and fire five hundred workers at the Imperial plant. People were not going to bite the hand that fed them. Fairbank employed fewer people in his bank, store, and oil farm at Oil Springs, so envious tongues could cut him to pieces without fear of too much reprisal.

By 1890, Petrolia was the oil capital of Canada, supplying nearly all of the country's oil needs, home grown. Her citizens were proud of the town's position in the oil world. And Petrolia's oil barons were building castles and making legends fit for kings. One such legendary event was the regal marriage of Jake Engelhart and Charlotte Eleanor Thompson.

In 1891, at forty-four years of age, Jake Engelhart ended his bachelor days. His marriage to Charlotte Eleanor Thompson, age twenty-eight, was a regal affair, heralded as "the most brilliant event in the annals of Petrolia's favored circle."[9] One hundred years later people still talk about *the* wedding. A young tour guide, showing visitors through Petrolia's historic well sites, recently gushed: "For *the* wedding, Mr.

Englehart gave the bridesmaids diamond rings. Would you believe? *Diamond* rings!"

In the evening, one day before the end of 1891, Charlotte walked down the aisle of Christ Anglican Church on the arm of her brother-in-law, George Moncrieff, Q.C., M.P. for East Lambton. Nellie and Bella Moncrieff, nieces of the bride, were bridesmaids, the lucky young women who received diamond rings from the groom. Irving Dittenhoefer, Englehart's nephew, a well-known New York city lawyer, was best man.

The *Petrolia Advertiser* reported:

The church was gorgeously decorated with plants and flowers and was crowded to overflowing. . . . Many of the guests were in full evening dress. . . . The bride's gown was a triumph of the Modistes Art — white faille, with court train trimmed with chiffon and orange blossoms. Her veil was pinned with a costly diamond fleur-de-lis, a gift from the groom. In her hand was a beautiful bouquet of lilies of the valley and orchids. The bridesmaids were attired in yellow bengaline and carried golden crooks decorated with roses.[10]

A reception was held at the elegant Moncrieff home where forty-nine telegrams of congratulation from all parts of North America awaited the couple. At midnight, the railway station was flooded with people pushing and shoving, hoping to get a close look at the bride and groom who were en route to New York and England for a honeymoon. The reporter, struggling to see the newlyweds, was only able to get a fleeting glimpse of Mrs. Englehart, "enveloped in a rich sable coat."[11]

Two days before the wedding, Englehart was invited to step into the old barreling room on the Imperial Oil premises. One hundred and fifty employees had gathered there to present him with wedding gifts — a handsome marble urn, mounted on a pedestal of onyx, imported from Paris, and an easy chair. "The cost had been subscribed to by every employee," the *Petrolia Advertiser* reported.[12]

Englehart had never taken a real vacation. When he traveled, there was always work involved. Now, his honeymoon offered him a chance to travel with an attractive, congenial companion, and, while seeing the world, to forget the oil business. He introduced Charlotte to his brother, sister, and their spouses in New York. He had kept in close touch with these relatives through the years; family ties meant a lot to him. In England, friends introduced him to the game of golf, and he became such an avid golfer that, when he came back to Petrolia, he built a golf course on the grounds of his new home.

And thus an attractive young woman became chatelaine of Glenview, the home set like a medieval castle on the outskirts of Petrolia. In the springtime, with the scent of apple blossoms fresh on the warm air, the chatelaine and her consort hosted garden parties on the terraced grounds. Oil rich people, many of them founders of the Imperial Oil Company, attended these parties. They were colorful events — the women in ankle-length, pastel brimmed hats, often carrying parasols in a myriad of colors. Once a year when the flowers at Glenview were in full bloom, Charlotte and Jake invited people to come and pick bouquets to place on the graves of loved ones in Hillsdale Cemetery. Petrolians came in great numbers for these "Cemetery Days." Often the flower beds were stripped bare, but it didn't matter. People counted more than flowers. Besides, gardeners would plant more seeds and bulbs, and within weeks the place would bloom again.

People knew from the tender way that Englehart touched his wife's hand, from the affectionate way he looked at her — often from the other side of the garden — that he loved her deeply. It was a love that would last for seventeen years.

DEATH OF THE FATHERS

While Petrolia celebrated the marriage of two of the oil establishment's foremost members, the city grieved the deaths of several of the industry's pioneers. In late 1890, the father of the oil industry in North America became ill in Hamilton. James Miller Williams, age seventy-two, with his wife at his side, lingered for two weeks at his estate, Mapleside, then on 26 November 1890, he died of congestion of the lungs.[1] With his death, the early chapters of petroleum discovery came to a close. Thirty-three years had elapsed since those precarious beginnings when Williams and Charles Nelson Tripp had tapped into the mother lode. Williams had quickly developed a commercial enterprise that had expanded slowly but surely into a many-faceted industry. No doubt he had been surprised at the growth and potential of this child he had sired. Certainly, he was one of the few who had become a multi-millionaire with oil money.

Other great oil barons followed the funeral procession to the Hamilton Cemetery, namely Jake Englehart, John Henry Fairbank, Frederick Fitzgerald, Benjamin VanTuyl, and George Moncrieff, but the ranks were now growing thin. Tripp, Williams, and Hugh Nixon Shaw had now passed on, and three years before Williams' death, Leonard Vaughn, at age fifty-two, had died suddenly in Petrolia; he was buried near Titusville where he was born and where his family's roots ran deep into Pennsylvania soil.

John Henry Fairbank did not seek another partner to replace Vaughn in his banking business; instead, he employed a manager. One historian commented that "the new arrangement clearly showed his dissatisfaction with the partnership form of enterprise."[2] After Vaughn's death, the bank's successes began to decline due to stiff competition from trust companies and chartered banks. Fairbank, still fiercely independent until the day he died many years later, refused to amalgamate with chartered banks, a move that might have saved his enterprise from the total collapse it experienced later. "They may think Vaughn & Fairbank not permanent," he once said to his son Charles, "but it is. It is a matter of pride. You can carry it on and John after you."[3]

In spite of the deaths of old friends and partners in the oil business and the slow decline of his bank, Fairbank kept active. He was proud of his oil wells, the VanTuyl and Fairbank store, and the Crown Savings and Loan Company that he founded with the help of Englehart. As the years reeled away, he became more conservative and cautious in business deals, and Englehart, as vice-president of the company, found himself "snorting with impatience."[4] Shortly after Fairbank's death in 1914, Englehart urged Fairbank's son to devote his energy to making this company prosper: "The Crown has been dormant, though in existence 32 years. It should be awakened, and I feel confident that, with the new manager, Egan, it should make good. And with Moncrieff, yourself and others, there is no earthly reason why the Crown should not be somewhere near the head, and not at the tail, as it has been and is."[5] Englehart became president of the company on Fairbank's death, but his "aggressive attitude does not seem to have been much more successful than Fairbank's caution," Edward Phelps explains. "In 1936, during the big depression, the Crown found it advisable to merge with the Industrial Mortgage and Trust Company, of Sarnia."[6]

STANDARD OIL RAIDS

In 1890, Imperial Oil celebrated its 10th anniversary quietly, with business as usual. Frederick Fitzgerald and Jake Englehart were, no doubt, pleased with the progress of this child they had helped to launch. Their sales to the Prairie provinces and British Columbia had increased rapidly through the years; in these far-flung outposts their agents sold directly to stores where their sturdy, practical five-gallon cans of kerosene were well-known. In fact, one hundred years later, a few of these historical, empty cans have been found on farms, in mines, and at lumber camps.

With the introduction of new industrial machinery in the 1880s and 1890s, Imperial developed lubricating products to meet those needs. By 1895 their price list included just over one hundred products and brands.[1] And as John S. Ewing reports, the company had branches in "London, Saint John, N.B., Kingston and Montreal; in the next four years Hamilton and Windsor were added, in addition to a considerable increase in the number of storage warehouses.... The Toronto branch of the Imperial Oil Company was... the Royal Oil Company. By 1896 the latter had storage warehouses at Peterborough,

Barrie, Orillia, Midland and Guelph.''[2]

After Imperial's 10th anniversary, Canada was hit by a depression. Although not as severe as the depression of 1873-74 that sent Petrolia drillers to foreign lands to earn a living, still it affected the oil industry. John S. Ewing in his history of the company notes that in ten years the population of Canada had increased only 11 percent, ''less than half the rate of growth in the United States over the same period. . . . Fortunately for Imperial, the Prairie regions and British Columbia had maintained rapid increase and so had sales there. But in the Maritimes, in Quebec, and in Ontario there were what amounted to mass migrations from eastern Canada south to the United States, from Ontario westward to the promise of rich, free farmlands. The replacement of sail and wood by steam and steel had robbed the Atlantic provinces of their position as mariners, the towns and rural areas of Quebec saw their people attracted by the industries of New England and New York.''[3]

There were several reasons why Imperial's rising star paused and began a downward spiral beginning in 1891. The recession and the relatively slow growth of the country was one reason, but another was the arrival of a dreaded competitor — John D. Rockefeller's Standard Oil Company. Late in the 1880s, this firm moved across the border, especially into the Maritimes where, in 1888, Standard Oil Company formed the Eastern Oil Company. In Ontario, Standard gradually bought out small refiners and negotiated low railway rates.

Rockefeller used four procedures to control a big percentage of refining operations in the United States: he amalgamated refineries in the main business areas; opened reliable agencies to purchase crude oil; cut costs; and employed marketing skills with great efficiency. He also used many devious tactics to attain these objectives. The one that got the best results consisted of putting pressures on competing refiners until they finally gave in to merger or acquisition with the big octopus. Standard was then in a position to bring under its wing the best of these failing competitors. Rockefeller would use such tactics to great advantage in Canada as he had done in the U.S.A.

Imperial was unable to control or buy out all small refiners who remained staunchly independent. And, although Imperial controlled fifty percent of the refining operations in the country, there was some erosion from within the company itself. Founders William Spencer and John Minhinnick cranked up their own refineries in the late 1880s, for example.

In 1889, Standard purchased the Bushnell Company Limited in Montreal, and through Bushnell, it bought a refinery in Sarnia, chiefly for a marketing site. ''With these inroads into the Canadian market,'' Charles Whipp and Edward Phelps observe, ''Standard suggested sharing business with Imperial, possessor of the largest refinery and marketing system in Canada. But, Imperial Oil rejected this 'friendly connection', and so Standard applied pressure through extension of its Bushnell operations across the country, by 1894 hitting hard at Imperial Oil's markets.''[4]

In order to compete with Standard, Imperial

needed more capital. The directors looked to Great Britain for this. Fitzgerald, writing to Frederick J. White, managing director of the Colonial Development Corporation in England, said that he was sending pro forma statements "showing what the Imperial Oil Company's profits would have been had the company sufficient capital to work with and make the necessary extensions and meet the requirements of the trade."[5] Negotiations for the sale of Imperial to the Colonial Development Corporation began at the end of 1895, and it isn't recorded whether Imperial first approached White on the sale, or whether White first approached Imperial. It is known, however, that White came to Petrolia and toured the oil fields; later, at the suggestion of Fitzgerald, the noted English petroleum geologist, Boverton Redwood, traveled to Petrolia to inspect the oil fields.[6]

Although Imperial gave a number of time extensions to the British firm, White seemed reluctant to make any firm commitment. No doubt the great swindle of British investors by the notorious Harry Prince some fifteen years before made White cautious. While the British firm played for time, Standard, in the guise of Bushnell Company Limited, bought into the Queen City Oil Company, controlled by Samuel Rogers, then the largest jobber and distributor of oil in Canada. At the same time, Bushnell took over the Fairbank-Rogers and Company refinery in Petrolia (cranked up a few years before by Fairbank and Samuel Rogers), and in 1897 bought out another inactive refinery, the Alpha, at Sarnia.[7]

When the octopus to the south sent its tentacles right into Imperial's back yard, Englehart and Fitzgerald knew they had to begin sharing business with a once-hated and feared competitor. While they struggled to maintain control of the Canadian oil industry, William H. McGarvey gained control of the Austro-Hungarian oil world, fending off all newcomers, even Standard Oil, and saw his daughter married to a member of the Austrian royal family of von Zeppelin.

ROYAL PETROLIAN

William H. McGarvey often made trips back to Petrolia to recruit drillers for work in Austria. It was not difficult to lure men away from the fields of Enniskillen, for hard oilers were natural adventurers, and a strike on the other side of the ocean offered not only adventure but big money. The drillers who worked for McGarvey enjoyed top wages, and those who returned to Petrolia came back wealthy men.

When returning to Petrolia in the 1890s, McGarvey learned of the troubles that Englehart and Fitzgerald were experiencing in trying to keep Imperial afloat, but he could only commiserate with them on these troubles. He had succeeded in building a petroleum empire that had made him the most important entrepreneur in Austria-Hungary; a caring man, he no doubt would like to have helped Englehart and Fitzgerald to save Imperial, but there was little he could do.

After drilling their wild well in 1883, Bergheim and McGarvey acquired enough capital to buy up hundreds of acres of land in the regions of Boryslaw and Krosno where they also built refineries. In 1895 they re-named their enterprise the Galician Karpathian Petroleum Company (Galizisch-Karpathischen Petroleum Aktien Gessellschaft), and built a huge refinery at Maryampole near the city of Gorlice in the southeast corner of Galicia. At this time they also bought up a number of small oil-producing and refining operations, and one of them was the Apollo Oil Company of Budapest. This sale made them famous not only in Austria, but also in Hungary. Bergheim maintained a home in England and traveled back and forth from the oil fields. McGarvey moved his family into an old regal palace in Gorlice.

Considered the biggest, most efficient enterprise in Austro-Hungary, the Maryampole refinery was built in six months. It employed one thousand men, mostly peasant farmers whose wives took care of farm work while their husbands were on the job. A small city eventually grew up around the refinery when the company built homes for workers so that they could be near the job site. Communal dining halls were also erected for the benefit of employees.

McGarvey recruited his two younger brothers, Albert and James in Petrolia, for work in Galicia, and he put them in charge of oil fields in the Maryampole

area. One day as James ate lunch in one of the plant's dining areas, a nervous employee shot him.[1] Fortunately, he recovered from his wounds and eventually returned to work, but the incident made Canadian field men uneasy. They were not used to supervising so many different, hostile nationalities in a country where pistol-shootings were often used to settle problems. At least one attempt had been made on the life of Emperor Francis Joseph, and his son, wife, and nephew all met violent ends.

About the time that James was shot, a cave-in occurred in the winding shaft of the nearby Stebnik salt works. A reporter conjectured that "this cave-in shows a development southeast of the Truskawiec spa in the Worotysseze brook — near the McGarvey drillings."[2] No doubt oil men working for McGarvey felt earth tremors from this cave-in, but other than that, there is no evidence of related geological problems in the oil field. But as in Petrolia, Rockefeller was on the prowl, trying to gobble up all big and little entrepreneurs. Nobel felt the Rockefeller presence in Baku. McGarvey felt it in Galicia where he formed a cartel with twelve major and several minor refiners to keep Rockefeller out. This only stopped the big octopus for a few years, then he swallowed up some of the refiners, but McGarvey remained a big independent until the First World War.

The Journal of the International Society of Drill Technicians and Drill Engineers reported that in 1880, before Bergheim and McGarvey arrived in Galicia, 30,000 tons of oil were processed in a year. By 1890, after this team had been in operation for seven years, their production alone amounted to 240,000 tons. By 1912 they had expanded to 1,800,000 tons.[3] Apart from the hundreds of wells in production and the refineries in operation, the company offered for sale oil field supplies patented by McGarvey — electric drills for dry or water drilling, Canadian drill rigs, steam engines, and steam boilers. Just before the turn of the century, McGarvey and Bergheim built another enormous refinery at Trieste, a harbor city on the Adriatic Sea. This seaport refinery gave them an advantageous position for shipping their oil products all over the world.

Emperor Franz Joseph was so pleased with McGarvey's exploratory work and subsequent production and refining of oil that he decorated him at a special ceremony in the Imperial palace in Vienna. Two editors summed up McGarvey's contribution to Austria as follows: "In 1882 McGarvey brought the Canadian drilling method to Galicia, pushing all the other methods into the background, all the more because the Canadians acted as so-called 'piece-workers'. This really promoted a faster upturn of oil mining. By means of the Canadian method, one could penetrate into greater depths. Hermetically closed pipes and water locks in the drilling holes were introduced at the same time and gave the oil extraction its efficient form."[4]

McGarvey moved into a palatial home in Vienna where he and his family were quickly accepted by the aristocracy. His children attended private schools and spoke German, Polish, and English fluently. Persons who were related to ruling kings in Europe were often visitors in his home.

By 1895, little May, born in Petrolia, was nineteen years of age, and that year Vienna papers announced her forthcoming marriage to Count Eberhard von Zeppelin, a brother of Count Ferdinand von Zeppelin who later invented the airship. The von Zeppelins acquired royal status in 1802 when King Friedrich raised Grandfather Ferdinand von Zeppelin to a Count and made him his Minister of Foreign Affairs. Ferdinand was subsequently decorated by the sovereigns of Austria, Russia, and Prussia for services rendered to these crowned heads of Europe.

The bridegroom and his brother grew up in a castle that stood on top of a hill overlooking a large lake. The whole estate was enclosed by a thick wall with two portals for the entrance. The wide driveway was lined with tall poplar trees, and a circular fountain with weeping willows graced the impressive approach. Lush vineyards and orchards were a pleasing part of the landscape within the walls. The bride and groom spent part of their honeymoon at this estate.

The wedding was held at high noon on November 12, in an old Gothic German Protestant Church in Vienna. The *Petrolia Advertiser* gave coverage of the event in two columns, with pictures of the bride and groom, and a picture of McGarvey, then fifty-two years of age. His dark hair was now touched with gray, but his moustache was still dark and full. He stood tall and wore a dark suit with white shirt and high, starched collar. The son of a storekeeper in Wyoming, Canada, William McGarvey looked every inch an aristocrat, wealthy, confident, capable of wielding enormous power. The editor of the paper summed him up in Shakespeare's immortal words uttered by Brutus: "There is a tide in the affairs of men, which, taken at the moment, leads on to fortune."[5]

The *Petrolia Advertiser* reported that the night before the wedding, McGarvey and his wife entertained at a gala reception held "in the princely apartments of the Grand Hotel in Vienna, after which a sumptuous supper was served in the palatial dining room — one of the finest dining halls in Europe. This was followed by dancing to music supplied by the world-famous Dreschner orchestra." The following day the bride and groom repeated wedding vows before Dr. Alfonso Witz-Stoeberg and Parson Gretzmacher, chairman of the Galician Protestant Church Assembly. "The bride was gowned in an elegant dress of moire silk trimmed with lace and orange blossoms. The bridesmaids were Miss Maggie Lake, daughter of Col. Lake, ex-president of the Incorporated Law Society of London, England; and Miss Kate McGarvey — May's young sister, born in Austria — who wore blue and white silk dresses. The groom was supported by Count Wisdehlen of the Garde Kuirassiers regiment; Baron Wachter of the Dragoon Regiment of Queen Olga; and Lieut. Wrede of the Uhlanen Konig William I regiment. All wore the gala uniforms of their respective regiments."[6] The bridegroom was a lieutenant in the Uhlanen Konig William I regiment, and for the wedding, he wore his uniform.

Following the service a breakfast reception was held at the Grand Hotel where seventy-two telegrams of congratulation from all parts of the world awaited the couple. The list of the guests read like a royal social register: Baron de Wolf-Sommersea, Captain of the Imperial Guard of the Czar of Russia; August von Gerayski, Member of the Austrian and Galician Parliament; Count Stefan Goetzendorf Grabowski; Baron Ferdinand Schoesberger; Baron Victor Schoesberger; Baron Stael von Holstein; Stanislaw Ritter von Syczepanowski, Member of the Austrian and Galician Parliament; Mrs. Arthur Lewis, sister of the celebrated actress Helen Terry. Bergheim and his wife were present, and Fred J. McGarvey, brother of the bride, along with hundreds of others. The bridegroom gave the bride a tiara of diamond stars to be worn on state occasions.[7]

With May's marriage, the McGarveys became an integral part of European royalty — with all the power, status, opulence, and problems that go with regal life.

William H. McGarvey became chief administrator for the Galician Karpathian Petroleum Company, working in Vienna. Often he traveled to job sites to confer with his partner John Bergheim, who, as field engineer, took care of all technical aspects of the vast petroleum empire. Just after the turn of the century, Bergheim was killed in a road accident in England, thus ending a long, amiable partnership, and McGarvey carried on alone.

William McGarvey was a brilliant salesman and businessman. His failures and successes in

Enniskillen had honed him into a sharp entrepreneur for the oil fields of Galicia. Like other early Canadian pioneers, he had learned to cope with the deadly fluctuations of oil prices and the cut-throat antics of producers and refiners. He and Fairbank were of one mind: united we stand, divided we fall. He formed early cartels of producers and refiners in Galicia and these certainly kept Standard away from him and helped to stabilize prices, too.

McGarvey was also a political animal. He had been Reeve of Petrolia on two different occasions and Warden of Lambton County, an appointment that broadened his scope in politics to the county level. He had also run as a Conservative candidate for East Lambton in a federal election. Although he did not win, old timers commented that he put on an impressive campaign. He was certainly aware of the rumblings of political discontent in Austria just prior to the First World War, but he probably did not view war as an outcome.

He was finely tuned to the changing petroleum industry. In Germany, as early as 1885, Karl Benz built a three-wheeled vehicle that had an internal combustion engine. A year later the strange-looking horseless carriage was driven through the streets of Munich to the surprise of everybody who saw it. By 1887, six years before Henry Ford invented his first automobile, a German engineer, Gottlieb Daimler, invented a four-cycle, single cylinder motor, and two years later a two-cylinder engine that gave greater power. In 1894 a Frenchman, Emille Levassor, produced his first automobile. These new inventions called for a fuel that was a volatile fraction distilled from crude oil.

As a relative and friend of Count Ferdinand von Zeppelin, McGarvey was, without a doubt, invited to view the finished airship invented by Ferdinand in 1900. This first zeppelin was 420 feet long, 38 feet in diameter, and had a hydrogen-gas capacity of 399,000 cubic feet. Steered by forward and aft rudders, the ship was driven by two 15-horse power Daimler internal combustion engines, each one rotating two propellers. McGarvey must have been very excited about this invention as well as the monoplanes that were beginning to appear in European skies. By 1906, Germans had adapted the diesel engine to submarines. McGarvey was aware that all of these inventions — the horseless carriage, airship, monoplane, and submarine — needed fuel, and he was processing close to two million tons of crude oil a year. The future looked to be even more promising than the past. It was an exciting time to be alive.

Great Britain and Germany were interested in converting their navy ships from coal to oil. McGarvey knew the statistics on this type of conversion: a ship that was fired with oil could go twice as far as one fired with coal before having to take on fuel again. Also, it took from one to three days to build up fires in a ship's boilers to make steam from coal, and it took that much time to load coal into a ship. Oil was faster to fire and load; also, it burned cleaner.

Rumors have floated around Petrolia for years that Winston Churchill, First Lord of the Admiralty in Great Britain, asked McGarvey to consult with

British Navy officials on the importance of converting ships to oil. There were very important oil men in Britain at that time, just prior to the First World War, and one of them was Sir Boverton Redwood, a geologist who was knighted a few years earlier by King Edward VII for his contribution to oil technology. Great Britain's biggest problem was that she did not have access to petroleum; Austria-Hungary and Germany were well supplied with McGarvey's oil fields, but Britain did not have any wealth like this at her back door. The manner in which she eventually acquired oil fields in Persia is another story. As early as 1913, four battleships being built in England were re-designed for oil.

A biographer in 1912 said that McGarvey is "the principal owner of the oil wells of Galicia which he still manages and controls; has also established many other industries depending for their existence on his company; employs 2000 men, many of his principal managers being Canadian." The same biographer quoted the *Montreal Star* which proclaimed that McGarvey was "a brilliant example of Canadian skill and enterprise."[8]

STANDARD TAKEOVER

Although Standard was often pictured as a demon, gobbling up everything in its path, it really became a savior to Imperial. "Not only had American enterprises managed to capture a third of the Canadian market but, for the first time, demand for oil began to exceed supply, and there was a pressing need for expansion," the authors of *The Story of Imperial Oil* explain. "It was not forthcoming in either Canada — then in a recession — or Britain, and Imperial looked for help south of the border."[1]

The struggle to sell the company began in 1895 when negotiations were begun with White in England. Not all the shareholders wanted to go this route, but Englehart and Fitzgerald were eventually able to convince them that selling the company was the best action to take. When the English firm dragged its heels, not wanting to make a firm commitment, Fitzgerald went to New York to meet with representatives of Standard Oil. "While he was there he wired Petrolia for more information on the company, and statements similar to those which had been sent to White were promptly dispatched," John S. Ewing reports. "Apparently, the material . . . presented was sufficient, and a preliminary agreement for sale was arrived at in that month [May 1898]."[2]

Standard bought seventy-five percent of the capital stock and inventories of the Imperial Oil Company, Limited, and shareholders decided that the working capital would be increased from half a million dollars to a cool million. Robert Page reports that "the net profits for the last full year of Imperial under Canadian management — 1897 — were 61,500 dollars; they rose under the new regime to 498,500 dollars in 1899 and 614,800 dollars in 1900."[3] These figures clearly show the potential of the Imperial operation when the directors had the necessary working capital.

The takeover was quiet and orderly. As Page notes, Standard avoided "a messy political fight which could combine all the rhetoric of the antitrust feelings with the anti-Americanism latent in Canada."[4] Standard retained the "Imperial" name, no doubt to keep the die-hard British happy. When Imperial was negotiating its sale to Standard, Sir Wilfrid Laurier was prime minister, and, although he was French Canadian, the country was still entrenched in British values. Certainly, Standard was well-advised to keep the takeover as quiet as possible and to avoid any anti-American demonstrations. However, the transition was not completely devoid of anti-American feelings.

A year before final papers were signed, Standard wanted changes in Canadian laws with regard to importing and selling oil products. Three changes were essential: a lower duty on imported oil; permission to import products in steamer tank ships; and permission to sell from horse-drawn tank wagons. As John T. Saywell reports in his history of White Rose, "such changes could only be secured through legislation. Thus the issue moved into the realm of politics, a realm in which Standard, from long experience in the United States, was very much at home."[5]

Tank wagons and tanker ships were not permitted in Canada because the government thought they were not safe. However, Englehart and Fitzgerald, well versed on changes within the industry, would have been the first to admit that transportation of oil products by bulk was far superior to transportation by cans and barrels. The duty on oil coming from the U.S.A. was six cents per Imperial gallon. Standard wanted this tariff dropped completely. The *Petrolia Advertiser*, opposed to Standard's tough lobbying tactics, wrote: "It is idle to argue that free oil will lessen the price to consumers. On the contrary, it would raise prices. The great and terrorizing monopoly — the Standard — would break down our barriers and get possession of our markets, but only to reap a golden harvest, and the consumer would find, too late, that instead of a reasonable price, they would have to pay anything the great octopus deemed fit to ask, and get a much inferior oil."[6]

Needless to say, most independent Canadian producers, refiners, and retailers were opposed to changes in the law to accommodate Standard, as Saywell explains:

John Fraser, the member of parliament for Petrolia, told Sir Wilfrid Laurier. . . that the proposed changes would ruin the Canadian industry. Canadians, he wrote bitterly, would be placed at the mercy of a large and powerful foreign corporation. Competition would be impossible, for Standard alone had access to American crude oil, alone had the money to buy and operate lake tankers, alone had the funds to build large fleets of horse-drawn tank wagons. Others joined in the protest. Laurier and the Minister of Finance were flooded with petitions, protests, appeals. In June 1897 the question moved into the House of Commons where a long and heated debate took place. But it was all so pointless. The government had made up its mind. . . . The duties were reduced and imports by tankers were permitted, although tank wagon delivery had to wait another two years.[7]

Independent producers, refiners, and the politicians who represented them did not seem to understand that Imperial needed to sell its company. The directors had struggled to interest a British firm in this sale, without success, and no doubt they were pleased with the deal they had struck with Standard. No doubt the government was aware of all of these transactions, and also knew that Standard, if not given concessions, could completely destroy Imperial. Standard had a growing market in all regions of Canada, possessed the Frasch patent to eliminate the sulphur from the oil, operated the most modern methods to transport petroleum products, and had the necessary capital to expand operations.

It took three-and-a-half years to completely close all deals concerning the takeover, and in the end, shareholders received $810,187.86, or $324 per share, with par value of $100 — and they were able to keep their twenty-five percent interest in the company — all very generous offerings, indeed.

The Petrolia Advertiser covered the story of the Standard takeover without denouncing "the octopus"; significantly, Standard increased its advertising for Imperial in local papers throughout 1898. But as Robert Page notes, the "real consequence of the takeover struck Petrolia with the announcement of the transfer of the refining facilities to Sarnia; it was normal practice for Standard to locate them on deep water locations. In 1897 and 1898 Standard acquired a number of other Canadian refineries, so its position of dominance by the end of 1900 was far greater than Imperial had ever been. Eastern, Bushnel, Queen City and the other Standard subsidiaries were all merged into Imperial to form the foundation of the present company. Englehart and Fitzgerald were retained with generous salaries under exclusive service contracts, but the effective management of the company swung over to officials from the Buffalo refinery and Barstow from New York."[8]

The Petrolia Oil Exchange ceased to exist in 1897. With Standard moving in, the Exchange was not needed any more. The brick Exchange Hall had been built twenty odd years before and stood like a proud sentinel in Victoria Park, the center of town. Thousands of bulls and bears in the oil business had marched in and out of this hall, including Fitzgerald and Englehart; fortunes had been made and lost here; prices had been fixed; squabbling had erupted and warring factions had met with swords drawn. But after the Imperial refinery was moved piece by piece to Sarnia, the old Oil Exchange Hall suddenly broke out in flames and burned to the ground, a symbolic, defiant gesture.

An artist's representation of the Imperial Oil Company refinery at Sarnia.

Petrolia, the one-time oil capital of the world, began to look like a ghost town. At one time there were six hundred wells producing in Petrolia, with many wild wells spewing out oil to the heavens. These gushers were gone now, replaced by saltwater, the same malady that had stopped the rivers of oil at Oil Springs. Wooden derricks became lonely skeletons guarding deserted wells; grass grew over and around jerker lines that had once hooted softly as they moved back and forth pumping the life-blood of the town; people left by the hundreds to seek work elsewhere. A few, who were fed up with oil, went to look for a different mother lode, gold in the Yukon.

GHOST TOWN

The Klondike gold rush held no appeal for John Henry Fairbank, Jake Englehart, and William H. McGarvey. They had found another bonanza that had catapulted them to fame and fortune. Fairbank and his wife, after moving into their castle in 1891, lived regally and entertained royal visitors. Earl Albert Henry Gray, born at St. James Palace, London, became Governor-General of Canada in 1904, but previous to this he was a guest of the Fairbanks at Sunnyside. Lord Henry George Lascelles, sixth Earl of Harewood, married Princess Mary, great-granddaughter of Queen Victoria. He, too, previous to his marriage and on his first visit to Canada, spent several days as a guest of the Fairbanks.[1] These were only two of the many royal persons that Edna and John Henry could claim as friends.

She, however, only enjoyed her royal status for five years. While visiting a health spa in Pasadena, California in March 1896, she died at age sixty-seven. There are no records in Pasadena or Petrolia as to the cause of death. In the Personal and Society Column of the Petrolia paper, March 12, a notice appeared: "Rev. Mr. Brookman will arrive from Toronto on Saturday to conduct the funeral of the late Mrs. Fairbank."[2]

In studying Edna by the standards of the western world nearly one hundred years later, it seems that she was an intelligent woman, brilliant in business and money matters. She operated a farm by herself for several years when her husband was struggling to get a toe-hold in a precarious new industry. When not working, she loved a party. At one of the famous assemblies, she laughed and danced the quadrille with VanTuyl. Fairbank, it seems, did not dance, although he attended these fashionable balls. He did not often go to church, either; he was raised a Baptist in Upper State New York, and is listed as "Congregationalist" on the 1881 Petrolia census. Edna, a devout Anglican, attended church regularly. However, her business sense was acute even in church dealings. She loaned a large amount of money to the Anglican church in

Petrolia for the construction of a new building. She was determined to build the biggest and the best church in Ontario. Several years later the church defaulted on payments because the congregation had dwindled. She asked that her loan be paid immediately, and in full, but the church could not do this. She threatened to foreclose. Fairbank stayed out of this mess, and in the end both Edna and the church arrived at an agreeable settlement.

In Edna's day, men were expected to excel in business outside the home, and women — no matter how talented or intelligent — were expected to walk in their husband's shadows. This must have been frustrating for Edna who seemed to want part of the limelight, and certainly a good share of her husband's time. Fairbank was so busy that he probably could not spend many hours with his wife and children. Edna's very existence lay on the shoulders of a man who, like all mortals, had feet of clay. No doubt her imaginary illnesses were cries to be noticed, to be loved. She was buried in Hillsdale Cemetery, beside her first-born son who had taken his own life fifteen years earlier.

Four years after Edna's death, Benjamin VanTuyl, age sixty, fell ill with meningitis, and complicated by diabetes — a disease he had suffered with for years — he died. Thus ended a rewarding partnership with Fairbank. The terms of that partnership had been hastily scribbled by both men on a piece of paper thirty-four years before, launching the VanTuyl and Fairbank hardware store. Under the able guidance of the then handsome ex-cavalry officer, the business had flourished. Today, this store is still doing a booming business, with old John Henry's great grandson, Charles O. Fairbank, at the helm.

VanTuyl was buried in Hillsdale Cemetery beside his first wife, Kate. A year after his death, George Moncrieff, at age fifty-nine, succumbed to a heart attack. The old pioneers were slowly thinning out.

The transfer of the Imperial refinery from Petrolia to Sarnia did not worry Fairbank. He had seen many refineries come and go. He was a self-made man, independently wealthy. At the turn of the century he had 485 oil wells in production that netted him $10,000 a year.[3] He became sole proprietor of the hardware store after VanTuyl's death, and the store was a lucrative source of income. By 1901, when Fairbank turned seventy, the bank was failing, but he stubbornly refused to amalgamate with any commercial bank to keep it solvent. At that time, too, his health was deteriorating. He did not have the amazing strength that had been his in earlier years. Also, he was going deaf. He continued to live on in the big house with various housekeepers and a retinue of servants. He often threw open his magnificent ballroom on the third floor for the famous assembly balls, and he continued to entertain an occasional royal visitor. At a later date in the 1950s, long after Fairbank's death, Guy Lombardo's famous orchestra played for private dances in the ballroom at Sunnyside.

Jake Englehart was retained by Standard at a handsome salary, but much of his power in the company had been lost with the takeover. No doubt he felt sad that he had been unable to keep Imperial as an independent company, but he was an astute businessman and knew in his own mind that amalgamation had been the right way to go. At the turn of the century, Englehart was fifty-three. Historian Michael Barnes has painted a vivid picture of him at this time:

Beneath a high-domed balding head, wire framed spectacles perch on a rather thick nose which is somewhat out of place in the otherwise finely chiseled features. The glasses do not conceal a steady, direct gaze. A black ribbon holds the pince-nez lenses and fixes an air of old world elegance. The neatly trimmed vandyke beard and thick mustache are largely gray. A white starched collar is too tight and narrow to harbor a tie. That slim piece of sober cloth descends bravely into its waistcoat shelter almost as an after thought. In all, a portrait of a man who, while elegant and correct of dress, was not always confined by clothing or appearance. There is that rare combination, philosopher and man of action together in one frame.[4]

There is no doubt that Charlotte and Jake wanted to have children of their own. He was fond of his nieces and nephews in New York, remembering them with gifts on birthdays, visiting them often. She had grown up from age ten in a large, closely-knit family. By the time they celebrated ten years of married life, no children had appeared. Of course, behind closed doors, his sexuality was questioned. Could he sire a child? Was he impotent? Sterile? Did he have homosexual leanings? This was certainly idle gossip. Englehart, from all appearances, was a healthy, virile male who loved women, and especially his wife. If they heard these gossipy rumblings — and no doubt they did — the Engleharts ignored them and carried on with everyday business.

Charlotte was a true philanthropist. She inherited the trait from George and Isabella Moncrieff who were involved in political, civic, and church causes that promoted human welfare. Although Englehart had a tendency toward philanthropic acts, like setting up compensation for workers injured in his refinery, his judgment was often marred by the need to be tough in the business world. Through the years, Charlotte's talent for supplying goodwill toward her fellowmen mellowed the toughness in Jake.

This was evident one hot summer day around the turn of the century. He loved his golf course. He often invited Glenview guests — royal visitors and prominent men in government and the oil industry — to join him in playing golf on his nine-hole course. Every spring the creek beside the Glenview golf course swelled with water and flooded part of Engelhart's land. He became quite exasperated with this spring run-off. He had built picturesque foot bridges over the creek, and sometimes the rushing flood would sweep the bridges off their moorings and make match sticks of them. One particular summer he decided to do something about these spring waters. He hired half a dozen men to install a retaining wall. A week went by. He walked to the far end of his property to see how the men were doing. They were all sitting under trees, smoking pipes, and cigarettes. Englehart frowned and bellowed: "Look here you fellows — I hired you to get this work done. I'm paying you good wages, and what are you doing? Loafing?"

The foreman on the job rose from his sitting position on the grass and faced him. "Jake — have you looked at the thermometer? It's a hundred in the shade. Goddam, this is the first time we've sat down since noon lunch, and that was four hours ago."

Englehart grunted, turned abruptly away and hurried back to the house. Ten minutes later he reappeared with cigarettes and tobacco.

"Look men, enjoy your break," he said, handing out his gifts.

One old timer said that his father, an employee at the Imperial refinery when it was still in Petrolia, took sick and died. "Jake came by the house and told my mother that my father had been one of the best workers he'd ever hired. There were no insurances or welfare handouts back then, but my mother got a cheque, regular as clock work, every week till she died. It came from Jake. He never forgot, even after he left Petrolia."[5]

Another Imperial worker added, "Jake liked things neat, well ordered. When he ran the field outside Petrolia. . . he had a duplicate of everything made, rigs, pumps, jerker lines — the whole thing — for 225 wells. If anything broke down he could switch over and be back in business in two hours. He had two men make pipe fencing with hand-carved fence posts to go around the whole property — all 200 acres of it."[6]

After Englehart left Imperial, he returned for a visit in his private railway car. The men at the refinery, then located in Sarnia, heard that he was coming. One worker commented: "Every piece of brass was cleaned so's you could see your face in it. There wasn't a speck of dirt or grease on the rigs. Every hangar was just so, every engine tuned, every blade of grass in place. Why, we even spread carpets that some of the foreign drillers had brought home from Persia — spread them on the rig floors. Jake liked things neat and clean."[7]

After the takeover, Englehart became more involved in community projects. He served as president of the Conservative Association in Petrolia, director of the Bank of Toronto, and governor of the University of Toronto. In 1896, he met Sir James Whitney who became leader of the Conservative party in Canada that year. Son of a blacksmith, Whitney worked hard to put himself through school. He was called to the bar in 1876 at age thirty-three. Twelve years later he was elected member of parliament for Dundas, Ontario; and in 1905 he broke a thirty-three year Liberal hold in the province to become premier. At a later date he would lure Englehart away from the oil fields of Enniskillen.

ENGLEHART'S RAILWAY

In 1905, newly-elected Premier James Whitney was having problems pushing a railroad through the claybelt region of northern Ontario from the lonely hamlet of New Liskeard to a post office-trading post called Cochrane, about 125 miles north.

The Temiskaming and Northern Ontario Railway had a long history of problems. In 1900 the Ontario government had kicked out $40,000 to survey parties to map out a 125-mile route for the railway from North Bay to Cochrane where it would connect with the National Transcontinental (forerunner of the Canadian National). The government anticipated that this rail line would help to settle the north where 300,000 square miles of wild, unchartered territory lay in wait.

The country was thick with fir, jackpine, spruce, and cedar trees; a thousand un-named lakes and swift-flowing rivers etched the lonely landscape. There were few settlers.

Private enterprise did not rise up in any form to take over this mammoth railroad engineering job, and the Ontario government reluctantly accepted the responsibility. By 1905, 138 miles of track had been laid from North Bay to New Liskeard. The chairman of the Temiskaming and Northern Ontario Railway was responsible for all aspects of punching this railroad through to its destination. Each chairman disliked the job so much that three were appointed and quit all in the same year. The part-time position paid only $5000 annually, and kept men away from their families for months at a time. The chairman worked with axe men who cut the right of way, tracklayers who struggled on behind the axemen, and engineers who built bridges. The responsibility was awesome. And then there was the country. Forest fires often raged without mercy, mosquitoes and blackflies rose in hordes in the spring and summer, blizzards cut across the country in winter.

It was well known that Englehart never turned down a friend in need. One day in the spring of 1906, he was asked to meet with the premier in his plush office in Toronto. And so began a new career for Jacob Lewis Englehart. At fifty-nine he became Chairman of the Temiskaming and Northern Ontario Railway, ready to push 400 miles of track through virgin wilderness, a track that would eventually join the National Transcontinental.

The news that the Engleharts were leaving Petrolia sent shock waves through the community. Men who had worked in the Silver Star Refinery, and later in the Imperial Oil Plant, felt betrayed. They knew that Englehart's first love was oil. How could he leave the oil region of Southwestern Ontario to go railroading?

Hard oilers did not love the railroads. They'd had problems with high tariffs and the politics of moving oil from one region to another. They were always trying to circumnavigate the railroads. Once they built a plank road to cut out that troublesome middleman. Later, they built pipelines for the same reason. And now one of the most dedicated of oil men was joining that foreign legion of railroaders.

Tongues began to wag. Englehart was in debt, in trouble with the law, on the verge of bankruptcy — and running away. Of course, none of these accusations was true. In the summer of 1906 a plush Pullman railroad car suddenly appeared at the Petrolia railroad station. There were only a few people on the platform to wave goodbye, for Englehart, a very private person, had not divulged his time of departure except to a few, close friends. This was not a spectacular, final farewell for the Engleharts. Their home address remained "Petrolia," and they would return on special occasions to enjoy family and friends and Glenview Castle, a home they left in care of trusted servants and gardeners.

Petrolia itself was changing. Oil fever had raged for thirty-two years, from the time that Benjamin King had struck it rich in 1866 to the time that Imperial had moved its refinery to Sarnia in 1898-99. During that time, Enniskillen Township had ignored the potential of its rich agricultural lands. Men leased farms, but dug for oil; if black gold did not appear, they moved on to other sites. If a well produced, then the "farmer" bought the farm, but in the frantic push for oil, he ignored the growing power of the soil.

Yet as early as 1867, men had found ways to drain the swampy Enniskillen land. They dug open ditches, inserted tile or ash wood drains to run off the water. When drained, the loamy clay produced bumper crops of wheat, hay, and corn; gradually, farmers brought in herds of beef cattle and milk cows. A few men turned to sheep farming. There is a delightful story concerning an oil-man-turned-farmer named Sam who went into sheep farming. When the lambs were born, the farmers cut off their tails for hygienic reasons. Sam, however, was worried about one lamb. The stub of his tail was not healing as it should. He hiked over to his neighbor's farm and explained his problem.

"That's easily fixed," said the neighbor. "You just dab some crude oil on the stub of the lamb's tail, and . . . like a miracle, it's cured."

Sam had an oil well on his land that gave him a barrel a day. He followed his friend's advice and touched the stub of the tail with crude. A few days later he visited his neighbor again.

"I'm in a lot of trouble with that oil cure you suggested," he said. "You see, the tail stub is healing, but the ewe won't feed the lamb."

"No problem," said the neighbor. "You dab oil on the snout of the ewe and she'll warm up t' that lamb like a sunflower to the sun."

And so it seemed that Enniskillen oil could cure in more ways than one.

Although many hard oilers turned to agriculture and supplemented the farm income with a few barrels of oil a day, there were Petrolians who felt lost without a big refinery churning out dollars in their midst. Three years after Imperial moved out, a charter was granted to the Canadian Oil Refining Company.

Canadian Oil Fields Limited pump house ("The Fitzgerald Rig") in 1903, now located on the grounds of The Petrolia Discovery.

The founders were six influential Canadians and three Americans. A Canadian company, it built a refinery in Petrolia that began operations in November 1901, refining 10,000 barrels of crude a week. Six years later, this company went bankrupt and it was bought out by an American enterprise, a similar fate that befell Imperial. The new company, known as the Canadian Oil Companies Limited, continued to operate in Petrolia. Later it became famous under the banner of "White Rose," with a white rose pictured on all its products.

Oil production in Petrolia was on the wane at this time. The output had dropped from close to a million barrels in 1894 to a third of that by 1910. Refineries were forced to import a full complement of American crude to keep going. The Petrolia oil boom was over.

Englehart returned again and again to the town where oil money had built Victorian mansions and oil barons lived like kings. He followed, with great interest, the changes in the industry. In fact, he would never forget oil. It was too much a part of him to ever ignore, but his new railroad job was demanding, and like all jobs that Englehart undertook, he threw himself into it, heart and soul.

In 1903, a railroad worker named LaRose threw his hammer at a fox. It missed the animal and hit a rock. When LaRose retrieved the hammer, he realized it had chipped off silver-bearing rock. This was the beginning of the silver strike just south of New Liskeard where a mining town named "Cobalt" evolved overnight. Prospectors flooded into the area. By October 1905, the T & NOR had brought in 99,000 tons of freight and 86,000 passengers.[1]

Englehart, knowing these statistics, became fired with the same fervor for railroad construction that he'd had for oil refining, and he used the same plan of attack: "Capable management, well-defined relations with employees, care for detail — from minor ordering to major engineering works."[2]

The plush Pullman or President's Car, with the Chief's flag flying at the front, became his business office and his home. It moved along the track behind the building crews. By 1907, with twin ribbons of steel being laid steadily northward from New Liskeard, Englehart was happy. But that year, Charlotte became ill. Rumors persist in Petrolia today that she had contracted tuberculosis. There is no proof that she did, but in the spring of 1908 she realized she was pregnant with the child that they both so desperately wanted. Charlotte was then forty-five years of age, and this was her first pregnancy.

In the next few months her health deteriorated quickly. On 31 October 1908, the President's Car was en route to Toronto to take her to hospital, and it paused at Gravenhurst, a pioneer town located half way between Toronto and North Bay. Charlotte must have known that she was dying, for here, on that day, she drew up her last will and testament, whereby Petrolia inherited a Victorian castle. In her will, she said:

The Canadian Oil Fields Limited bull wheel pumping mechanism was the largest in the world when installed in 1903.

After the death of my husband and subject to his power and right to dispose thereof or otherwise deal therewith . . . I give devise and bequeath the residence occupied by my husband and myself with all the lands, buildings and appurtenances connected used and enjoyed therewith including the golf links property now owned by my said husband and by his will devised to me and particularly described in the conveyance thereof, to the Municipal corporation of the Town of Petrolia for the purpose of a General Hospital for said Town. . . that the said Hospital shall be called 'The Charlotte Eleanor Englehart Hospital.'[3]

Charlotte died on New Year's Eve that year, of complications due to her pregnancy. She was buried in Petrolia.

The year that Charlotte died, Henry Tripp, age eighty, living in Lynn, Massachusets, also died. He survived his brilliant, eccentric brother, Charles Nelson Tripp, by forty-two years, and will go down in history as the man who alerted his brother to the discovery of the gum beds in Enniskillen. Henry will always be fondly remembered in Schenectady as an itinerant photographer who had a talent for capturing the early life of the city on film.

There was no mention of Henry's death in Oil Springs or Petrolia where, in the early 1860s, he bought and sold land at a feverish pace. Although Englehart must have known Henry, and certainly would have heard of Charles Nelson Tripp, he probably did not know of Henry's death. And in 1908, he had his own grief with which to cope. As though he knew that the only antidote for a broken heart was work, he returned early in the spring to Northern Ontario to push more track through the vast wilderness.

In the summer of 1911, gold was discovered 60 miles north of New Liskeard. The rush that followed built up the town now called Kirkland Lake, but by that time, the railroad had passed Kirkland Lake and had reached Porquis Junction, about 50 miles north. Building this section was slow, torturous work, especially when the engineers had to span rivers like the Blanche River that turned and twisted at the bottom of a wide, rugged valley. Today, Kapkigiwan Provincial Park, part of this valley, lures travelers to waterfalls and granite rocks that rise awesomely above the river, but before the railroad bridge was built, settlers north of the valley felt cut off from the world. On the edge of the gorge, just north of the bridge, a town sprang up which settlers called "Englehart" to honor the man who was responsible for laying down the steel that opened up the north.

The lure of the wild and the fascination of the vast, lone land drew Englehart to the north. As a keen man of business, he also saw great possibilities for development of this big country, and he made other people feel his strong convictions. Speaking to the Toronto Board of Trade in the spring of 1911, he said: "Is it not your duty to assist in opening up that great northland, and to see that the settlers turn back

I, CHARLOTTE ELEANOR ENGLEHART, of the Town of Petrolea
in the County of Lambton, wife of Jacob Lewis Englehart of
the same place, Esquire, hereby make and publish this my last
Will and Testament, hereby revoking all former testamentary
dispositions by me at any time heretofore made.

I DIRECT my executors and trustees hereinafter named or
the survivors or survivor of them to set apart and (subject
to the power of sale hereinafter contained) to retain two
hundred shares in the capital stock of The Crown Savings &
Loan Company of Petrolea of the par value of fifty dollars
per share and out of the annual dividends or income thereon
as same is received to pay to my sister Kate Thompson Four
hundred dollars per annum and to my nieces Helen Glen Moncrieff
and Isabel Glen Moncrieff each the sum of One hundred and
fifty dollars per annum until the death of my said sister.
In the event of the death during the lifetime of my said
sister of either of my said nieces leaving issue her surviving
I DIRECT my executors to pay the income to which such niece
would be entitled if living to the issue of said niece In
the event of the death during the lifetime of my said sister
of either of my said nieces without leaving issue her surviving
I DIRECT my executors to pay to the survivor of my said two
nieces three hundred dollars per annum. In the event of the
income from dividends or other income from said stock not
amounting at any time to Seven hundred dollars per annum
I DIRECT my executors to make up the deficiency out of the
balance of my estate and on the death of my said sister I
DIRECT MY EXECUTORS to divide such shares or the proceeds
thereof, if sold, equally among my nephews and nieces, the
five children of the late George Moncrieff; AND I DIRECT
my executors as soon as may be after my death to divide my

or shares sold for all purposes relating to my will and shall
be dealt with accordingly.

My executors and trustees shall not be responsible
in case of loss by reason of their not selling such shares
or any of them should they depreciate in value.

I GIVE DEVISE AND BEQUEATH the residence occupied by my
husband and myself with all the lands, plant, buildings
and appurtenances connected, used and enjoyed therewith
situated in the Town of Petrolea or adjoining said Town to my
dear husband Jacob Lewis Englehart to have hold and fully
enjoy as if he were the absolute owner thereof for and during
his life.

After the death of my said husband and subject to his
power and right to dispose thereof or otherwise deal therewith
as hereinafter is provided I GIVE DEVISE AND BEQUEATH the
said residence property with all the lands buildings and
erections plant machinery and appliances with the appurtenances
thereon or in any way connected used and enjoyed therewith
including the golf links property now owned by my said husband
and by his will devised to me and particularly described in
the conveyance thereof, to the Municipal Corporation of the
Town of Petrolea for the purposes of a General Hospital for
said Town. PROVIDED and this devise is made on the express
condition that the Council of said Town within one year
after the death of myself and my husband or the survivor of us
duly passes a proper by-law of said Town accepting such devise
and entering into an agreement and covenant with my executors
and trustees or the survivors or survivor of them (which
by-law and agreement is to be satisfactory to my said
executors and trustees or the survivors or survivor of them)
to the effect that the Corporation of the said Town will within
such time as shall be agreed upon with my executors and
trustees establish equip and properly maintain and operate

*Last Will and Testament of
Charlotte Eleanor Englehart,
2 March 1909.*

*Advertisement for the Canadian
Oil Refining Company.*

from the trek of the West to the trek of the North —
to hold our people in our own back yard. The
province of Ontario with the northland at its back —
the Temiskaming country — is a whole Dominion in
itself."[4]

Without Charlotte, the big house in Petrolia
seemed to grow emptier year after year. Christmases
were especially lonely, and after Christmas in 1911,
Englehart gave Glenview to the town for a hospital.
On January 31st, a turnover ceremony was held in
his castle with people flooding through to see this
manor where a lord and his lady once lived. That
summer, Englehart officially opened the thirteen-
bed hospital. A young boy attended the event and
left this portrait of Engelhart: "After the ceremony, I
wandered out into the garden. Mr. Englehart was sitting all alone on the bridge with his head in his
hands. He was crying and when he saw me he asked
where my mother was. I said I didn't know. He took
me by the hand and we went back up to the big
house to find her. I remember not wanting to let go
of him because he seemed so lonely."[5]

Later, Englehart financed the addition of two
wings to the hospital, an operating room and a
nurses' residence. As one of his biographers notes,
"he equipped the maternity ward personally, and
provided in his will for the purchase of X-ray equipment. He endowed the hospital with four hundred
shares of Imperial Oil stock . . . and thus ensured its
independence and financial security for all time."[6]

Oil Refinery, Petrolea, Ont.

Site of the Canadian Oil Refinery at Petrolia in 1906.

THE END OF AN ERA

John Henry Fairbank, age eighty-three, died quietly at Sunnyside on 10 February 1914, four months before the outbreak of the Second World War. Records at the Hillsdale Cemetery office say he died of old age. His son Charles was no doubt at his bedside when he died. Charles had graduated with his degree in medicine several years earlier, and was his father's personal physician.

The *Petrolia Advertiser* reported:

Tuesday morning the end came to a life connected with the Canadian Oil industry from the beginning, with the death of John H. Fairbank, at his home here. . . . Deceased had been confined to the house for nearly two years and his death had been expected for some time. . . . Disease did not play a part in the dissolution; he weakened day by day and finally when all strength was gone, his spirit took its flight.

For nearly 50 years the late J. H. Fairbank led a busy life in the oil fields of Canada. . . . The "ups and downs" of the oil business were a disturbing element in his varied business interests, which frequently required much thought and guiding in order to avoid the rocks; he was always at the helm, with a clear head and quick action in times of trouble, and he came through it all with very few scars and a princely pile of this world's goods to

his credit. To the casual acquaintance he appeared cold and reserved, but such was not his nature when business affairs permitted him to be his natural self.[1]

Fairbank's estate was worth nearly $800,000 "after providing for cash bequests to his grandchildren and many of his employees," his biographer Edward Phelps notes. "Fairbank had directed that the remainder be divided equally between his two surviving children, Dr. Charles O. Fairbank and May (who had married in 1896 in Pasadena, California to a London man named Huron Rock). Since most of the holdings consisted of real estate, oil properties, and businesses which could not be readily liquidated because of economic circumstances at the time of the owner's death, the estate of J. H. Fairbank continued to function as a corporate entity."[2]

The funeral procession wound slowly from his palatial home to Christ Church and then on to Hillsdale Cemetery. Flags flew at half mast; places of business were closed. The Petrolia band played the *Death March* to mark the passing of this man, one of the last of the hard oilers who had mucked for gold in the oil-soaked swamps of Enniskillen.

Jake Englehart, listed as "Chairman of the Ontario Railway Commission," was among many special guests at the funeral.[3] He had known Fairbank for forty-eight years, since he, Englehart, had been a teenager, walking the muddy fields of Oil Springs looking for crude to feed his fledgling refinery in London. The two men had not always agreed with one another on business and oil matters, but even when they disagreed, they had not held grudges.

Four months after Fairbank's death, Archduke Francis-Ferdinand, heir-presumptive to the throne of Austria-Hungary, was assassinated by a Serbian nationalist. On 1 August 1914 Emperor Franz Josep h's sad message was published in newspapers throughout the land: "To My Peoples! — It was my dearest wish to devote the years which God in His mercy may still grant me to working for peace and to protecting my peoples from the heavy sacrifices and burdens of war. Providence has seen fit to decide otherwise."[4] The emperor was then eighty-four years of age. He died two years later and was, therefore, spared the experience of seeing Austria's defeat in 1918, and the end of the Habsburg monarchy.

Seven months before the assassination of the archduke, when the world was still at peace, William H. McGarvey celebrated his seventieth birthday. It was the 27th of November 1913. *The Journal of the International Society of Drill Technicians and Drill Engineers*, an Austrian magazine of that era, devoted a two-page spread to him and his accomplishments:

The life story of McGarvey would be incomplete without pointing to the success of his company. From a modest beginning, the Galizisch Karpathieschen Petroleum Aktien Gesellscheft (the Galician Karpathian Petroleum Company) got larger, till now, it is well into all of the Austrian Oil Industry. This year's drilling program includes 35 deep drillings. Today's crude oil production is monthly 12,000 tons. Besides the large crude oil output in Galicia and the refineries in Maryampol, the company built two refineries in Urgam, Hungary. So, he was showing that his company was busy in both parts of Austria-Hungary. Moreover, with his pipeline storage and transportation, building and selling drilling machinery, he was well known and supplied all of Europe and the whole world with his material.

In conclusion we want to mention that McGarvey was not an optimist. He looked at his drilling with the thought that — you can only say you succeeded after the oil is out of the ground. The oil industry would be in better shape if we had more men like McGarvey. He was adventurous in his work, always looking for new fields. He made Galicia the big oil country. To see how enterprising he was, we have to remind you of Markop. The "Deep Drilling Company" was — we are quite certain — established to make Markop the largest Russian oil field.[5]

Eight months after McGarvey's birthday, the world was at war. His oldest child, Nellie, had married a Vienna judge a few years earlier, and the judge, along with von Zeppelin, were pressed into military service for Austria. Many Canadians, working in the oil fields, hurried to England or Canada to enlist in the armed forces; later they would fight against Austria-Hungary in a war that was supposed to end all wars.

McGarvey was always proud of his Canadian birth. He considered Canada his homeland and little Petrolia, especially, the brightest spot in that homeland. His parents had moved to London in the early 1890s, and there they celebrated fifty years of married life. At that time, they bragged that four of their sons were working in the oil fields of Austria.

The demand for oil had changed from those early years. In 1879, Thomas Alva Edison held a public exhibition in the United States and had displayed his new invention, the incandescent electric light bulb. Three years later, Edison developed and installed a central electric power station in New York city, the first in the world. Kerosene light quickly paled to the light supplied by electricity. Then in 1893, Henry Ford built his first automobile, and in 1903 he opened his auto plant in Detroit to manufacture horseless carriages that used a fuel which the early oil men had discarded as useless — gasoline — a volatile fraction distilled from crude oil.

The years prior to the First World War were exciting, profitable years for McGarvey. In spite of the invention of electric lights there was still a big demand for kerosene in Europe and the demand for gasoline began to rise on the European market. McGarvey's oil fields and refineries had never been busier, but once war broke out in early August 1914,

the tide turned. In late August, Russian soldiers advanced into Galicia, and in the subsequent battles between Russians and Austrians, McGarvey's refineries were blown to pieces and his oil fields set ablaze. Many of the Canadian workers who did not leave when war was first declared were shot to death or taken prisoner by Russian troops. There are stories floating around Petrolia today that tell of daring escapes by a few lucky Canadian families who struggled through enemy lines and over mountains to get to England.

By the time that November rolled around, McGarvey was like a dying monarch, his crown toppling, his kingdom in ruins. People who knew him at that time told of a gaunt skeleton of a man, grieving for his suffering people — those from his beloved Canada, and those from his beloved adopted country, Austria. On his birthday in November 1914 he died of a stroke. He was buried in Vienna.[6]

Jake Englehart was grief-stricken when news of McGarvey's death filtered through to Canada. He had known McGarvey for as long as he'd known Fairbank, and his memory no doubt flashed back to the blue-eyed, good-looking young man he'd first met, the Reeve of Petrolia who wore elegant, stylish clothes and traveled around in a "dandy" buggy; the oil producer, refiner, and explorer who'd gone west with geologists to try to find a mother lode on another front. Perhaps he recalled McGarvey's deep voice with its slight trace of Irish accent, the lilting, often intense speech of a man who loved the oil business,

W. H. Mac Garwey †.

An seinem 71. Geburtstage traf ein Schlaganfall unser langjähriges Mitglied W. H. Mac G a r w e y, der dessen Tod herbeiführte.

W. H. Mac Garwey ist Kanadier, er hatte aber die Absicht, mit seinem, ihm in den Tod schon lange vorausgegangenen Geschäftsfreund Bergheim, seine Bohrmethoden persönlich in Europa einzuführen.

Zu diesem Zwecke unternahmen beide Genannten in Hannover eine Reihe von Bohrungen nach kanadischem System. War es nun der Umstand, dass die für europäische Gebirgsverhältnisse noch nicht angepassten Werkzeuge den gewünschten Erfolg vermissen liessen, war es die Einsicht, dass hier bedeutende Erfolge — nach amerikanischen Begriffen — nicht zu finden sind, kurz nach einiger Zeit übersiedelten die beiden Freunde nach Galizien, um hier die Verhältnisse zu studieren und die vorhandenen Oelfelder rationell aufzuschliessen.

Gabs ja auch in Galizien Schwierigkeiten, so hatte die Firma Glück und es wurden reiche Positionen erschlossen. Anfangs ging man nur bis 200—300 Meter, da man aber später darauf kam, dass reiche Oelhorizonte tiefer liegen, wurde das gebrachte System immer weiter vervollkommt und zum Schluss ein spezifisch „Galizisches Bohrsystem" herausgebildet, das auch erhöhten Anforderungen gerecht wurde.

Wir finden heute Bohrungen nach diesem System vielfach über 1000 Meter tief in die reichen Oelsande hinuntergestossen, die es der österreichischen Industrie ermöglichte —seinerzeit ein grosses Importland für amerikanische und russische Ware — selbst in steigender Weise zu exportieren.

An diesem Aufschwung der galizischen Erdölindustrie hat der Verstorbene in hervorragender Weise teilgenommen und als er der sich immer mehrenden Arbeit nicht mehr Herr wurde, übertrug er seine Rechte an die Galizische Karpathen-Petroleum-Ges. vorm. Bergheim und Mac Garwey. Diese mit reichen Mitteln arbeitende Gesellschaft erweiterte die Unternehmungen. Die grosse Maschinenfabrik lieferte nicht nur den eigenen Bedarf an Werkzeugen, sondern lieferte zum Teil den Bedarf für die übrigen Unternehmungen, sie hatte auch starken Export ins Ausland. Die Bohrunternehmung hat ihrerseits gleichfalls Abteilungen im Auslande im Betriebe, so dass die Geschäfte immer umfangreicher wurden. Früher leitete W. H. Mac Garwey den Grubenbetrieb selbst, in den letzten Jahren stand er der Wiener Direktion vor, die er bis in die letzte Zeit mit starker Hand leitete.

Galiziens Erdölindustrie, dessen Grossindustrie sein Werk bildet, wird ihm stets ein treues Andenken wahren.

An seine Stelle tritt nun sein Sohn F. J. Mac Garwey, der schon in jungen Jahren an dem Unternehmen sich beteiligte und für die Folge dasselbe weiter führen wird.

Nach Befreiung Galiziens von der Russeninvasion wird ihm dort ein reiches Feld der Arbeit erwarten, da wird er Gelegenheit finden, zu zeigen, dass er der rechte Sohn dieses bedeutenden Vaters ist. _____ H. U.

Obituary notice for William H. McGarvey as it appeared in an Austrian magazine.

a man who, through that business, had entered the powerful, dazzling world of royalty.

Englehart knew what it felt like to lose an empire; he had lost his own empire, Imperial Oil, to John D. Rockefeller. Englehart no doubt thought about the intense suffering endured by his friend, the heart-wrenching pain that probably contributed to his death. In fact, people in Petrolia today will tell strangers that McGarvey died of a broken heart.

The war was hard on everyone. Englehart found it difficult to recruit workers for the railroad because so many of the eligible young men had gone to war. Some of them left their picks and shovels by the railroad track and marched off to enlist. In spite of these hardships he pushed on with characteristic determination.

The Toronto *Globe* noted that "the chairman of the Temiskaming Railway Commission scrutinizes every item of expenditure, no matter how small, and signs every check and voucher which leaves the office, but beyond the municipality of small things he has the larger vision of the man who can plan and carry to fruition vast and vital projects. The visitor to the offices of the Commission finds its head easy of access, urbane of manner, debonair of person and always genial and obliging. It does not matter whether it be a member of the government who drops in, a railway magnate come to confer upon momentous matters, a newspaperman in search of information, or the humblest employee of the road with a grievance or a request — all alike are made to feel welcome and at ease."[7]

One historian, who has studied Fairbank and Englehart, was not as kind in his comments: "Englehart was a hard-driving son-of-a-bitch to the very end."[8]

The war years were tough on Engelhart. In 1914 when the first shots were fired, Englehart turned sixty-seven; and by 1919, after struggling with a shortage of workers, a shortage of steel, and the ever present menace of forest fires, he was tired. He resigned from his railroad job and retired to the Queen's Hotel in Toronto which stood on the site of the present Royal York. After Glenview was given to Petrolia for a hospital, he had moved to Toronto. Following retirement, he spent winters in California, returning to Canada when spring was in the air.

From his hotel room, he often wrote letters to his nephews, nieces, and friends, and there, too, he wrote his last will and testament: "To Lizzie Frank Englehart of New York, my five-stone diamond ring; to my nieces Blanche R. D. Dittenhoefer and Estelle Dittenhoefer of New York my French concave glass cabinet and contents; to my wife's nieces Eleanor Lyle and Nora Lyle, photographs of my late wife, her mother and myself; to my nephew Howard Lindeman my cobalt blue faience vase and two steel engravings, the wedding ring and the soap bubble. . . ."[9] His will named 119 beneficiaries. For many there were large sums of money, but also, each was given a sentimental article, perhaps a silver match box, a walking stick with a root handle, a mahogany desk. He remembered what each person had admired and left him or her that particular memento.[10]

In April 1921, while writing a letter to a friend, Englehart was stricken with a hemorrhage of the brain. Shortly afterward he lapsed into unconsciousness and died. He was seventy-four. The *Imperial Oil Review* reported that "with the passing of Mr. Englehart goes the last connecting link between the old and the new management of our company. Mr. Englehart, working with other important Ontario men, obtained the charter for our Company in 1880 and was closely identified with its fortunes since. As vice-president and director until his death, he was the sole surviving member of the original incorporation who continued to occupy a seat on the board."[11]

A special train left Toronto carrying the body for burial in his home town. Officials of two railroads — the T & NOR and the Grand Trunk — were on the train, along with prominent men from Imperial Oil and the government. In Petrolia, flags flew at half mast, shops were closed, window shades were drawn in respect. Hundreds of people lined the streets as the funeral procession moved from Christ Anglican church to the cemetery.

With Englehart's death, the final chapter closed on the history of the early petroleum days. He was the last of that hardy breed of men who scrambled through mud and slush to find crude oil for a hungry still. He had come, a penniless teenager, struggling with laundered money to set up a lucrative refining business. Hardships, at times,

seemed insurmountable, but he had toughed it out and had stayed. When he died, he was worth three and a half million dollars.[12]

Another founding father, Frederick Ardiel Fitzgerald, did not fare as well. He severed all connections with Imperial at the turn of the century, and by 1901 he hit hard times and was nearly bankrupt. "He owed Fairbank $6,500 on some oil-producing property and was obliged to surrender the land in lieu of paying the debt," Edward Phelps reports. "He wrote to Fairbank (from London), 'I am exceedingly sorry to have to say; — that matters with me are so mixed up here, that it is quite out of my power to take up your claim, in fact, at present I am perfectly helpless in the matter and obliged to abandon. If there is anything that I can do to assist in giving you immediate possession, I will only be too glad to do what I can. I regret very much that this property is thrown on your hands in this way, and grieved to have to be a party to it. I sincerely trust you may be able to handle it without a loss.' "[13] The firm of F. A. Fitzgerald & Co., Oil Refiners of London, became insolvent in 1902.

Three years after Englehart's funeral, Fitzgerald, age eighty-four, died quietly in London. The *Evening Free Press* made note of his passing with a front page story under the headline, "Pioneer of Oil Industry Dead." A picture showed a man who looked much like the Fitzgerald of early Imperial days, but his long beard was very white. He was buried in London, quietly, without fan-fare. And so a great era in the history of the Canadian petroleum industry — indeed an important chapter in the history of international business — came to an end nearly three score and ten years after James Miller Williams first struck oil in Enniskillen and the rock poured out "rivers of oil."

AFTERWORD

Today, many visitors feel a haunting loneliness as they wander around Oil Springs. Summer winds sigh through the wild grasses that cover the pioneer oil field where, more than a century ago, thousands of adventurers mucked for big stakes. The field — two miles square — looks diminutive and sad, with its broken fences, scattered pieces of ancient machinery, and three pole derricks white-washed with time.

The road to the railhead in Wyoming — the old mudhole that swallowed up horses and wagons — is now paved, and a sign alerts visitors: OIL SPRINGS, population 720. The main street of the village, which was once covered with a two-inch thickness of white oak, is also paved. Residents will tell interested tourists that, when re-paved a few years ago, several of the old oak planks were found underneath the rubble. Residents will also explain that there are still billions of barrels of oil locked within the confines of the earth beneath their feet. They hope that some day, somebody will find the key to unlock this bonanza, and then oil will again gush forth in Oil Springs, as it did in Hugh Nixon Shaw's day. Many attempts have been made through the years to release this "trapped" oil, but none brought back the famous gushers.

The main street of the village is quiet with few mementoes of those long-ago days when horse-drawn omnibuses clattered from one end to the other, when nine hotels did a roaring business, when saloon keepers pushed booze across counters in a steady stream of mugs, when blacksmiths' anvils rang through the din of street life, when prostitutes added a colorful touch of sin. A two-story Masonic Temple, built in 1866, is a solemn reminder that there was a down-to-earth, conservative side to life in those wild days before the scare of the Fenian Raids cleaned out Oil Springs. A blacksmith's shop stands, weather-beaten, on one corner. It is rumored that, in 1862, the blacksmith here sharpened the drill bit that Hugh Nixon Shaw used to drill his wild well. East of the village a 300 acre oil field operates without seismograph crews, computers, or any of the dazzle of the 1990s. On this property, there is a mural covering one side of a barn that shows a brightly-colored tank wagon pulled by horses; bold letters proclaim: 1861, FAIRBANK OIL.

Charles Oliver Fairbank, fifty-one, owner-operator of Charles Fairbank Oil Properties Limited, says his oil business is the only one in the world that has been operated continuously by the same family since 1861. After Charlie's great grandfather John Henry Fairbank died in 1914, his grandfather operated this field, then his father, and now he is at the helm.

From a path that runs beside two large earth storage tanks used in his great-grandfather's day, visitors can see, in the distance, a circular dip in the land, the only mark left of Canada's first wild well drilled by Hugh Nixon Shaw in 1862. Although Shaw lost his life in this well, nothing today reminds people of that tragedy or the excitement of the gusher, except a few holes in the ground, now covered with grass, the remains of clay storage tanks.

In Charlie's oil field, four hundred wells are being pumped, all connected by 7½ miles of jerker lines that squeak and hoot softly as they move back and forth. Walking beams nod up and down above the wells as they draw oil to the surface. Various electric pumps power the jerker rods today, but in John Henry Fairbank's day, horses were used for power, and later, steam engines.

There is a pungent odor of oil here with a distinct smell of sulphur. In the old boom days, oil men told visitors that this skunky odor was "the smell of money." Charlie says it is not the smell of money for him. He's called a "stripper," or marginal producer, and it's very costly to maintain a small field like his. He breaks even, cost-wise, but in order to do this, he and his crew of six workmen must improvise and apply band-aid techniques of repair to antiquated equipment. A high-school teacher by trade, he has learned to use the forge where he fixes or makes from scrap metal the hangars that support the jerker lines.

Around Oil Springs, there is a good collection of machinery left from the boom of the 1860s and the deeper drillings of the 1800s. Charlie, himself, has found a number of old relics, including a drilling rig and part of a steam engine. And a modern museum marks the spot where, in 1858, James Miller Williams, assisted by Charles Nelson Tripp, tapped into an underground vein of oil, launching the petroleum industry in North America. Visitors can see the old well site with the three-pole derrick made of ash wood used by Williams and early oil men to lower and lift their drilling tools. A spring-pole drilling rig is set up above the well as it was in Williams' day.

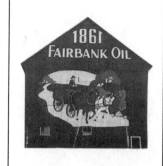

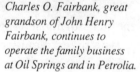

Charles O. Fairbank, great grandson of John Henry Fairbank, continues to operate the family business at Oil Springs and in Petrolia.

Field displays show tank wagons and jerker rods pumping oil from distant wells. Inside the main museum building, there are exhibits of early machinery and oil well supplies, along with pictures and videos that tell the story of the booms and busts that put Oil Springs on the map.

Nine miles north of the village lies the town of Petrolia. With a population of 4,600, the city has a delightful Victorian-age look in its churches and the upper-story lines of its old downtown buildings. Modeled after Gothic and Tudor revival architecture of the 1880s and 90s, the homes of early oil barons still grace the streets. On the main road overlooking Bear Creek, John Henry Fairbank's beautiful mansion, finished in 1890, is still there, but it has been sold and re-sold through the years and now is in need of repair.

The Englehart house, willed to the town as a hospital, has had modern wings added through the years, but the old Victorian architecture still peaks out from these modern facades. Originally an eight-bed hospital in 1911 when opened, it now has 72 beds, and minor surgeries are performed there.

Victoria Hall, finished in 1889, is an elegant old girl that Petrolians are proud of. A tragedy occurred in 1989, shortly after the celebration of its centennial; the hall caught fire and flames licked high into the night with sparks blowing wildly in the wind. Hours later when firefighters brought the fire under control, only the outside brick skeleton was left standing. Petrolians felt devastated. Sadly, they assessed the damages and cost of repair. A year later, they rallied support for a six million dollar reconstruction project that kept the outside skeleton intact and carefully copied the old, inside architecture. The hall stands today, as elegantly as it did yesterday, near a park on the main street. Professional theatrical performances are on-going there in winter and summer.

Several small industries are based in Petrolia today, but many people earn a living by commuting to Sarnia 17 miles away, where they work in the industries spread out along the St. Clair River. However, die-hard Petrolians do not like the big city to the northwest. The trauma of 1898-99, when the Imperial Refinery moved lock, stock, and barrel from Petrolia to Sarnia, still rankles like a raw nerve. In fact, some people blame Sarnia for the demise of the halcyon days of oil in Petrolia. Other people say that the dislike of the city is simply the small guy's inherent jealousy of the big guy.

Petrolians are immensely proud of their oil heritage. In fact, there is a tendency for them to forget that Oil Springs was the first to discover oil. Although a few wells had been drilled as early as 1862, the boom did not come to Petrolia until Benjamin King struck his gusher in 1866 and the Fenian Raids cleaned out Oil Springs. Thousands of wells were punched in a 20 square mile radius — ten times the size of the pioneer field at Oil Springs. Many of these old wells are still there; they have been a source of trouble in the past when people with children wanted them capped. Hundreds have been plugged, and others have been fenced off, but people still say there are old, gaping holes in the ground. Helen Corey, whose husband Harrison "Tip" Corey spent a lifetime in the oil business, speaks of an area that she calls "the Devil's Half Acre" located on the edge of town where wells were drilled so close together that a person could spit from one to the other. Many years ago, a twelve-year-old girl fell into one of these oil wells in the Devil's Half Acre, and she was drowned.

Although the town fathers insist that there are no open wells or pits, one old timer calls this statement "hogwash". She says the town fathers do not know where the wells and storage tanks were located. Early oil men did not leave any blue prints, and today grass and weeds are growing over these old sites. She points to The Petrolia Discovery site, a 60-acre tourist attraction, where sixteen old wells are working in a 19th century pioneer oil field. Hundreds of visitors are ushered through this area every year where a guide explains the workings of the machinery, the history of the wells, and the story of the early oil barons. It is a popular spot for tourists, but the old timer says The Discovery site is not fenced in, and every year vandals tramp all over it and steal costly equipment.

Until twenty years ago, an oil well was operating in the back yard of a town church located on a busy corner of the main street. During services, the pump squeaked and squawked while drawing oil from the ground. Ministers gave their homilies with little notice of this thumping in their back yard.

In farmers' fields throughout Enniskillen Township, it is common to see lone old wells, the cross beams nodding up and down, drawing a barrel of crude a day or a week, for the farmer. This, along with flocks of sheep that are prevalent here, give a pastoral ambience to Enniskillen, which somehow seems to fit.

Imperial Oil has grown from those early, precarious years in Petrolia when there were 500 employees. Today, Canada-wide, Imperial employs close to 10,000 people. Frederick Ardiel Fitzgerald and Jacob Lewis Englehart would, no doubt, be delighted if they could see the picture of their offspring today. Six refineries are in operation, nationwide: Ioco, Norman Wells, Strathcona, Sarnia, Nanticoke, and Dartmouth; four coastal and Great Lakes tankers are busy throughout the shipping season. The Products' Division markets more than 700 petroleum products throughout Canada under well-known brand names, notably ESSO. The motoring public is served through 3,700 retail outlets.

In Englehart's day, kerosene and paraffin wax were the big selling products, but by 1907 the motor car changed this picture. *The Story of Imperial Oil*, a short history published by the company in 1991, gives an interesting anecdote:

The first motorists bought gasoline in cans or open buckets from company warehouses or from grocery or hardware stores where they normally purchased kerosene. Then, one day in 1907, one of the sputtering horseless carriages pulled up to Imperial's Vancouver warehouse to refuel alongside the horse-drawn tank wagons. Suddenly, there was a loud bang; it sounded as if a pistol had been fired. Actually, it was only the sound of the car backfiring, but the place was in chaos. Horses reared and teamsters cursed. When the fuss died down, the warehouse foreman banished motor cars forever. The next day, Imperial's Vancouver manager, C. M. Rolston, started Canada's (and quite possibly North America's) first service station outside the warehouse at Smithe and Cambie streets. It was a three-sided shed, opened to the street; a converted 60 litre hot water tank, painted red, dispensed Imperial gasoline to cars at the curb, through a garden hose.[1]

The demand for gasoline revolutionized the refineries of the day; suddenly, they had to process a greater proportion of crude oil into gasoline to meet the new demands, quite the reverse of refining procedures in the time of Fitgerald and Englehart.

In 1903, there were 178 cars in Canada; in 1910, there were 6000; by 1920 that figure had climbed to a quarter of a million horseless carriages. The First World War, with its demand for fuel for army lorries, aircraft and tanks, pushed Imperial to rebuild its Sarnia refinery, which became the largest in Canada, capable of processing 10,000 barrels of crude a day.

In 1915, British Columbia's first refinery was opened at Burrard Inlet, east of Vancouver; in 1916, refineries in Regina and Montreal appeared; and in 1923, a refinery at Calgary began operations. These plants served well for many years before they were

closed to accommodate more modern facilities.

The Second World War put a strain on Canada's fledgling oil industry. Many ships of Imperial's tanker fleet went to war. Seventy-seven Imperial men died when their ships were sunk by German U-boats.

Following the oil booms in Oil Springs and Petrolia, no new fields were discovered until gas was found at Turner Valley, Alberta in 1914. Then, in 1920, Imperial drillers found oil at Norman Wells, Northwest Territories. A refinery was built there, the most northerly refinery in the world, and it became high profile in 1944 when a pipeline was built from Norman Wells across 450 miles of virgin forest and mountains to Whitehorse, Yukon. A refinery was hastily built in Whitehorse, and it processed oil from Norman Wells, supplying Alaska Highway construction crews and military aircraft during the crucial days of the war. In 1947, the Yukon refinery was dismantled and moved, piece by piece, down the Alaska Highway to Dawson Creek, where the mammoth parts were loaded on railway cars and sent to Edmonton for re-assembly.

Following the war, there was a great need for oil in North America. The Sarnia refinery had, earlier, been linked by pipeline to mid-continental oil fields in the U.S.A., but the demand outran the supply. Between 1919 and 1946, Imperial drilled 133 wells in Western Canada — all dry. The company was discouraged, and in 1946 had to be coaxed by exploratory crews to drill a series of last chance holes outside of Leduc, Alberta. It was worth the effort to go that extra mile: on 3 February 1947, Leduc Number One blew in with great excitement, dumping oil into the surrounding countryside before it was controlled. The subsequent field was a spectacular success, producing more than 200 million barrels of crude in the next decade. The Whitehorse refinery, re-assembled that year in Edmonton, processed this oil until 1975 when four Imperial refineries (at Edmonton, Regina, Winnipeg, and Calgary) were closed and replaced by the modern Strathcona refinery, the largest in Western Canada.

One operation in which Imperial has been involved would cause Englehart and Fitzgerald to shake their heads in disbelief. Pioneer oil men and refiners dumped unwanted by-products of oil into creeks and surrounding fields. They probably did not even consider that this procedure was bad for their health and damaging to the very earth itself. As early as 1918, Imperial, concerned with water pollution, installed a large, modern separator in its refinery at Dartmouth, Nova Scotia, to extract oil from the sewer water before it was discharged into the harbor. And in 1924, the company hired Dr. R. K. Stratford to form a research department at the Sarnia refinery. Stratford bought a handsome home on the river south of the city, and he piped drinking water into his house directly from the river and chlorinated it himself. He liked to entertain, and on one occasion, so the story goes, his guests complained that the water he was using was spoiling the taste of their whiskey. Stratford began to investigate the river water.

Phenol was being used in a major process which

Oil wagon on the grounds of The Petrolia Discovery.

Imperial employed for refining lubricating oils. Stratford found that there was phenol in the river water and it was the culprit that spoiled good whiskey, and tea and coffee, too. As a result of this find, he helped form an organization called the St. Clair River Research Committee to study river pollution. In the late 1950s, Dow Chemical Company in Midland, Michigan developed a micro-biological process for the removal of phenol from plant discharge water, and Imperial, along with other industries associated with the St. Clair River Research Committee, adopted this process. By 1967, this committee became incorporated under the name of the Lambton Industrial Society, and sixteen petro-chemical industries in the Sarnia area joined in an effort to clean up the environment.

A program to reduce hazardous wastes at the company's chemical operations in Sarnia, initiated in 1981, resulted in a ninety-three percent reduction of liquid hazardous wastes and a seventy-five percent reduction of solid hazardous wastes by the end of 1990.

Protection of the environment is on-going, and, although some people say that Imperial did not move soon enough on such concerns, they are unaware of early efforts. The company provides financial support for environmental research and has, for example, helped to establish the Environmental Research Center at Trent University, Peterborough, Ontario.

In 1989, Imperial purchased Texaco Canada Inc. at a cost of five billion dollars. It was said to be the most important event for the company since the discovery of oil at Leduc. Texaco's refinery at Nanticoke, Ontario — one of the most modern and efficient in North America — became an excellent partner for Imperial's refinery in nearby Sarnia. With the acquisition of Texaco Canada, Imperial consolidated its position as the leading Canadian petroleum company. But the 1990s remain a challenging decade. Because of the recession of 1990-92, Imperial cut its work force from 15,000 to 10,000 across country. In Sarnia alone, the number of employees were dropped from 1800 to 1500.

The challenge of the 1990s comes with the need to be the best player in a market where demand is stagnant but excess production capacity exists. Gasoline consumption is the same as it was fifteen years ago, jet fuel demand has not changed, and the demand for diesel fuel has grown. Although there are more vehicles on the road now, they have a better burn of fuel through improved technology, so the demand has not expanded correspondingly.

Imperial Oil chairman Bob Peterson said at a press conference in Sarnia recently that staff cuts created a "pain that lasts" within the company, but people recognize that it is a changing world which requires adaptability. There will be a continual search for ways to be more productive and the result will be a more intense, more focused corporation with a more stable future. Jake Englehart and his fellow founders of this company and indeed this industry would no doubt apply their considerable ingenuity and tenacity to this search, if they were alive today.

NOTES

FOREWORD

[1] R. Bruce Harkness, Letters to Ernest C. Miller, May to July 1960, Dr. Douglas Jones File, Oil Museum of Canada, Oil Springs, Ontario.

THE HARD OILERS OF ENNISKILLEN

A Brief History of Oil

[1] Charles K. Johnson, Interview with Hope Morritt, 1968, Walpole Island Indian Reserve, Walpole Island, Ontario.

Gum Beds

[1] Jean Turnbull Elford, *A History of Lambton County* (Sarnia: Lambton Historical Society, 1967), p. 77.

[2] The Marchioness of Dufferin and Ava, *My Canadian Journal* (London: John Murray, 1891), p. 343.

[3] Abstract Book E, Registry Division of Lambton County, Lands Records Office, Sarnia, Ontario.

[4] Jean Turnbull Elford, *Canada West's Last Frontier: A History of Lambton* (Sarnia: Lambton Historical Society, 1982), p. 44.

[5] Eric Jonasson, *The Canadian Geological Handbook* (Winnipeg: Wheatfield Press, 1976), p. 44.

[6] Elford, *Last Frontier*, p. 44.

[7] L.P. Hurd, Letter to Lewis Burwell, 13 July 1832, Geological Survey of Canada Library, Ottawa, Ontario.

[8] Lewis Burwell, *Report on Enniskillen Survey*, 9 January 1833, Geological Survey of Canada Library, Ottawa, Ontario.

[9] I.C. McClusky, "The Evaluation of Oil Refining," *Imperial Oil Review*, Imperial Oil Archives, Toronto, p. 34.

[10] R.B. Harkness, "Ontario's Part in the Petroleum Industry," *Canadian Oil and Gas Industries* (February/March 1951), p. 1.

[11] Harkness, "Ontario's Part," pp. 1-2.

Charles Nelson Tripp

[1] *Canada Directory*, Village of Bath, 1850.

[2] "The Original Oil Man of Canada," *The Huron Signal*, 21 September 1866.

[3] Death Notice, Charles Nelson Tripp, George Smith Collection, Lambton Historical Room, Lambton County Library, Wyoming, Ontario.

[4] Harkness, "Ontario's Part," p. 31.

[5] *Lambton County Atlas*, 1880.

[6] Enniskillen Abstract Books, Registry Division of Lambton County.

[7] Enniskillen Abstract Books.

[8] Edward Phelps, "John Henry Fairbank of Petrolia (1831 – 1914): A Canadian Entrepreneur," M.A. Thesis, University of Western Ontario, 1965, p. 20.

[9] G. A. Purdy, *Petroleum: Prehistoric to Petrochemicals* (Toronto: Copp Clark, 1957), p. 22.

[10] John S. Ewing, "The History of Imperial Oil Limited," Unpublished Study for Business History Foundation Inc., Harvard Business School, Imperial Oil Archives, Toronto, 4 vols., vol. 1, p. 23.

[11] Harkness, "Ontario's Part," p. 31.

[12] Ewing, vol. 1, p. 22.

[13] City of Hamilton Census, 1851.

[14] Thomas B. Wilson and Emily S. Wilson, *Directory of the Province of Ontario 1857* (Lambertville, New Jersey: Hunterdon House, 1987).

Paris International Exhibition

[1] "The Paris Exposition," *The Weekly Spectator* (Hamilton), 17 August 1855.

[2] "The Paris Exposition," *The Paris Monitor,* 6 August 1855.

[3] *The Weekly Spectator*, 30 June 1855.

[4] "The Paris Exhibition," *The Daily Globe* (Toronto), 22 August 1855.

[5] *The Weekly Spectator*, 20 December 1855.

[6] J. C. Tache, M.P., Secretary to the Committee of the House of Commons, "Canada at the Universal Exhibition (Paris) 1855" (Toronto: Queen's Printer, 1856), pp. 41, 48, 155, 372, 395.

[7] R. J. Seckel, Le Conservateur Responsable de la Salle des Catalogues et des Bibliographies, Bibliotheque Nationale, Paris, France, Letter to Hope Morritt, 21 May 1992.

[8] Harkness, "Ontario's Part," p. 31.

[9] Harkness, "Ontario's Part," p. 32.

The First Oil Well

[1] Tripp File, George Smith Collection, Lambton Historical Room, Lambton County Library Headquarters, Wyoming, Ontario.

[2] Harkness, "Ontario's Part," p. 31.

[3] Book # TpB 373, Lots 16 & 17 Concession 2, Enniskillen Township, Registry Division of Lambton County, Lands Registry Office, Sarnia, Ontario.

[4] Robert Garnham vs Charles N. Tripp, Court of Common Pleas, Toronto, Ontario, February 1857.

[5] Tripp File, George Smith Collection.

[6] Harkness, Dr. Douglas Jones File.

[7] Fergus Cronin, "North America's Father of Oil," *Imperial Oil Review* (June 1958), p. 23.

[8] Cronin, p. 23.

[9] J.F. Caley, "Paleozoic Geology of the Windsor-Sarnia Area, Ontario," Canada Department of Mines and Resources, Geological Survey, Memoir 240, King's Printer, 1946.

[10] Tripp File, George Smith Collection.

[11] *Sarnia Canadian Observer*, 22 July 1857.

[12] *Harkness*, "Ontario's Part," p. 16.

James Miller Williams

[1] Hamilton Census, 1851.

[2] *Hamilton Evening Times*, 26 November 1890.

[3] John G. Taylor to Hope Morritt, 23 July 1991. John G. Taylor, the great grandson of James Miller Williams, provided this family history.

[4] Cronin, p. 23.

[5] Taylor to Morritt, 23 July 1991.

[6] "Important Discovery in the Township of Enniskillen" *Sarnia and Lambton Observer Advertiser*, 5 August 1858.

[7] *Sarnia and Lambton Observer Advertiser*, November 1858.

[8] Neil F. Morrison, Ph.D., *London Free Press*, Summer 1859.

[9] Elford, *Last Frontier*, pp. 125-26.

[10] Assessment Roll, Enniskillen Township, 1859, Lambton Historical Room, Lambton County Library Headquarters, Wyoming, Ontario.

Jigging Down

[1] *Daily Spectator* (Hamilton), 12 June 1860.

[2] "From Wentworth Mission to St. Andrew's Church 1856-1935, Church History," Williams' File, Oil Museum of Canada, Oil Springs, Ontario.

[3] Max W. Ball, *This Fascinating Oil Business* (New York: The Bobbs-Merrill Company, 1940), p. 334.

[4] Ewing, vol. 1, p. 33.

[5] Charles Fairbank to Hope Morritt, 3 March 1992.

[6] *Daily Globe*, 26 August 1861.

[7] Harkness, Dr. Douglas Jones' File.

[8] Cronin, p. 24.

[9] Cronin, p. 24.

[10] Writ to Sheriff, Historical Room, Lambton County Library Headquarters, Wyoming, Ontario.

[11] R. B. Harkness, Sir William Logan's G.S.C. Statistics, 1865, Letter to Ernest C. Miller, 10 June 1960.

[12] Enniskillen Abstract Books.

[13] London Census Records, Reel C1097, Middlesex County, 1861.

[14] R. Galbraith Letter, Microfilm Reel, Enniskillen Historical Records, Lambton County Library Headquarters, Wyoming, Ontario.

[15] Larry Hart, "Pioneer Lensman," *Gazette* (Schenectady, N. Y.), 17 September 1974.

Hugh Nixon Shaw

[1] J. Harvey Johnston, edited by Arthur B. Johnston, *Recollections of Oil Drilling at Oil Springs, Ontario* (Tillsonburg, Ontario: Harvey F. Johnston, 1938), p. 3.

[2] *Leader* (Toronto), 29 March 1861.

[3] Ben Fiber, "Oil Pioneer's Grandson Attends Centennial," *Sarnia Observer*, 2 July 1958.

[4] William Perkins Bull, *Spadunk or From Pragmatism to Davenport United* (Toronto: George J. McLeod Ltd., 1935), p. 90.

[5] Jacqui Brown, County of Huron Museum and Historic Gaol, Letter to Hope Morritt, 16 August 1991.

[6] Tremaine Map of Peel County, 1859.

[7] Winnifred Shaw, Letter to Hope Morritt, 29 May 1991.

[8] J. S., "An Admonitory Occasion," *Christian Guardian*, 31 May 1865, p. 86.

[9] "Enniskillen Oil Region," *Leader* (Toronto), September 1861.

[10] Bull, pp. 60-61.

[11] Harkness, "Ontario's Part," p. 33.

[12] John Henry Fairbank, Letter to Edna Fairbank, Oil Springs, Ontario, 16 March 1862, Historical Room, Lambton County Library Headquarters, Wyoming, Ontario.

John Henry Fairbank

[1] Phelps, p. 4.

[2] Phelps, p. 5

[3] Phelps, p. 10

[4] Phelps, p. 21

[5] Phelps, pp. 22–23

[6] John Henry Fairbank, Letter to Edna Fairbank, 15 October 1861, Historical Room, Lambton County Library Headquarters.

[7] Phelps, p. 22

[8] John Henry Fairbank, Letter to Edna Fairbank, 15 October 1861.

[9] "Mortality in Hamilton," *Daily Spectator* (Hamilton), 15 January 1858.

[10] John Henry Fairbank, Letter to Edna Crysler, Galt, Ontario, 15 April 1858, Historical Room, Lambton County Library.

[11] Phil Morningstar to Hope Morritt, 14 June 1991.

[12] "Radway's Ready Relief," *Weekly Spectator*, 18 October 1855.

[13] J. H. Fairbank, Diary, 18 November 1863, Historical Room, Lambton County Library.

[14] Fairbank, Diary, 18 November 1863.

[15] M. O. Hammond, "The Oil Rush to Enniskillen," *Imperial Oil Review* (September 1929), p. 3.

[17] Ball, p. 34.

Wild Wells
[1] Bull, pp. 90-93.

[2] "The Oil Wells of Enniskillen," *London Free Press*, 14 February 1862.

[3] Michael O'Meara, *Oil Springs: The Birthplace of the Oil Industry in North America* (Oil Springs, Ontario: Oil Museum of Canada, 1963), p. 9.

[4] "The Oil Springs," *Globe* (Toronto), 20 February 1862.

[5] *Christian Guardian*, 19 February 1862, p. 30.

[6] Selwyn P. Griffin, "Petrolia, Cradle of Oil Drillers," *Imperial Oil Review*, Imperial Oil Archives, Toronto, p. 20.

[7] Tom Evoy to Hope Morritt, 27 May 1991.

[8] Ewing, vol. 2, p. 5.

Race Riot
[1] Johnston, p. 8.

[2] "Disgraceful Riot at Oil Springs," *Sarnia Observer Advertiser*, 20 March 1863, p. 2.

3 Martin Luther King Jr., *Conscience for Change*, CBC Massey Lectures (Toronto: Canadian Broadcasting Corporation).

[4] Charles Watten, "Racial War Waged in Early Sixties," *Petrolia Advertiser Topic*, 30 January 1936.

[5] "Disgraceful Riot," p. 2.

[6] S. G. Howe, *The Refugees from Slavery in Canada West: Report to the Freedmen's Enquiry Commission* (Boston: Wright & Potter, Printers, 1864), p. 43.

[8] Watten, "Racial War."

[9] Phil Morningstar to Hope Morritt, 14 June 1991.

Canadian Oil Association
[1] Phelps, p. 31.

[2] Fairbank, Diary, 18 October 1862.

[3] Fairbank, Diary, 13 November 1862.

[4] Elford, *A History of Lambton County*, p. 40.

[5] Bruce A. Macdonald to Hope Morritt, 15 June 1965.

[6] Victor Lauriston, *Blue Flame of Service: A History of Union Gas Company and the Natural Gas Industry in Southwestern Ontario* (Chatham, Ontario: Union Gas Company of Canada Limited, 1961), p. 2.

[7] Lauriston, *Blue Flame*, p. 2.

[8] "Historical Sketch of the County of Lambton," *Illustrated Atlas of the Dominion of Canada* (Ottawa: Queen's Printer, 1880), p. x.

[9] *Sarnia Observer and Lambton Advertiser*, 14 February 1862.

[10] Fairbank Letter to Edna Fairbank, 16 March 1862.

[11] Johnston, p. 6.

Royal Honors
[1] "International Exhibition: The Awards," *Globe* (Toronto), 25 July 1862.

[2] "Canada at the Great Exhibition," *Globe* (Toronto), 8 July 1862.

[3] *Globe* (Toronto), 25 July 1862.

[4] "Oil News," *Globe* (Toronto), 23 July 1862.

[5] Ewing, vol. 1, p. 50.

[6] *Globe* (Toronto), 12 September 1861.

[7] "From the Oil Springs," *Globe* (Toronto), 18 February 1863.

[8] Fairbank, Diary, 11 February 1863.

[9] "Mr. Hugh Nixon Shaw of Cooksville," *Christian Guardian*, 27 May 1863,. p. 88.

[10] J. S., "An Admonitory Occasion," *Christian Guardian*, 31 May 1865, p. 86.

[11] J.S., "An Admonitory Occasion," p. 86.

Old Fairbank
[1] Fairbank, Letter to Edna Fairbank, 31 January 1864.

[2] Phelps, pp. 27–28.

[3] Fairbank, Letter to Edna, 31 January 1864.

[4] Phelps, p. 33.

[5] Fairbank, Letter to Edna, 18 April 1862.

[6] Edna Fairbank, Letter to J. H. Fairbank, 18 January 1863.

[7] Fairbank, Letter to Edna, 3 December 1865.

[8] Fairbank, Letter to Edna, 3 December 1865.

[9] Phelps, p. 36.

Drilling and Refining Technology
[1] Purdy, p. 26.

[2] Phelps, p. 61.

[3] Harkness, Dr. Douglas Jones' File, p. 158.

[4] Ball, p. 213.

[5] *Footpaths to Freeways: The Story of Ontario's Roads* (Toronto: Historical Committee & Public Information Br., Ontario Government, 1984), p. 37.

[6] John Maclean, "The Town That Rocked the Oil Cradle," *Imperial Oil Review* (June 1955), p. 9.

[7] Phelps, pp. 25-26.

[8] Wanda Pratt and Phil Morningstar, *Early Development of Oil Technology* (Oil Springs, Ontario: Oil Museum of Canada, 1987), p. 12.

[9] Pratt and Morningstar, p. 12.

[10] Dianne Newell, *Technology on the Frontier: Mining in Old Ontario* (Vancouver: University of British Columbia Press, 1986), p. 37.

[11] Pratt and Morningstar, p. 11.

[12] Pratt and Morningstar, p. 11.

Jake Englehart and William H. McGarvey
[1] Hugh M. Grant, "Capital Accumulation in the Early Canadian Petroleum Industry, 1857 – 1880," M.A. Thesis, Department of Economics, University of Winnipeg, p. 7.

[2] Grant, p. 7.

[3] Grant, p. 8.

[4] Grant, p. 8.

[5] Grant, p. 9.

[6] Ian Sclanders, "The Amazing Jake Englehart," *Imperial Oil Review* (September 1955), p. 4.

[7] Sclanders, p. 4.

[8] Elford, *A History of Lambton County*, p. 47.

[9] *Globe* (Toronto), 12 September 1861.

The Return of Charles Nelson Tripp
[1] Lauriston, *Blue Flame*, p. 3.

[2] Johnston, p. 17.

[3] Book #G6029, Indenture 11 Aug. 1866, Registry Division of Lambton County, Lands Records Office, Sarnia, Ontario.

[4] *Leader* (Toronto), copied from *The Huron Signal*, 21 September 1866.

[5] Phil Morningstar to Hope Morritt, 14 June 1991.

[6] Charles Nelson Tripp Obituary, Tripp File, Lambton Historical Room, Lambton County Library Headquarters.

[7] Book #K7775, Indenture 25 Oct. 1870, Lands Records Office, Sarnia, Ontario.

Boom and Bust
[1] Victor Ross, *Petroleum in Canada* (Toronto: Southam Press Limited, 1917), p. 34.

[2] Ross, p. 34.

[3] Ball, p. 318.

[4] Ross, pp. 35–36.

[5] Selwyn P. Griffin, ''Petrolia, Cradle of Drillers,'' *Imperial Oil Review* (1930), p. 21.

[6] Griffin, p. 21.

[7] Bruce A. Macdonald to Hope Morritt, 15 June 1965.

[8] Elford, *Last Frontier*, p. 147.

[9] Bruce A. Macdonald to Hope Morritt, 15 June 1965.

[10] Tom Evoy to Hope Morritt, 17 May 1991.

[11] Harkness, Dr. Douglas Jones' File, p. 33.

[12] Lauriston, pp. 182-183.

[13] Rollie Hall, Superintendent, Canadian Eastern Division, Haliburton Services Ltd., to Hope Morritt, 1970.

[14] Harkness, Dr. Douglas Jones' File, p. 41.

THE OIL BARONS OF PETROLIA

The Petrolia Discovery

[1] Maclean, p. 8.

[2] Victor Lauriston, *Lambton County's Hundred Years: 1849 – 1949* (Sarnia: Haines Frontier Printing), p. 171.

[3] Maclean, p. 9.

[4] Phelps, p. 54.

[5] Lauriston, *Lambton*, p. 169.

[6] Maclean, p. 9.

[7] Lauriston, *Lambton*, p. 169.

[8] Phelps, p. 51.

[9] John Henry Fairbank Obituary, *Petrolia Advertiser*, 11 February 1914.

[10] Charles Whipp and Edward Phelps, *Petrolia: 1866 – 1966* (Petrolia: The Advertiser-Topic and the Petrolia Centennial Committee, 1966), p. 1.

[11] *Globe* (Toronto), 20 April 1865.

[12] Grant, p. 16.

[13] Grant, p. 21.

[14] Jacob Lewis Englehart, Biography, Englehart File, Historical Room, Lambton County Library Headquarters, Wyoming, Ontario, p. 1.

[15] Grant, p. 10.

[16] Englehart Biography, p. 2.

[17] Englehart Biography, p. 2

[18] Dorothy Bedggood, ''Brigden's Founder Colorful Adventurer,'' *The Gazette* (Sarnia), 29 May 1974.

[19] Bedggood, ''Brigden's Founder.''

[20] Phelps, p. 59.

[21] Phelps, p. 59.

[22] Phelps, p. 59.

[23] Phelps, pp. 60–61.

VanTuyl and Fairbank and Vaughn

[1] Charles O. Fairbank to Hope Morritt, 4 May 1992.

[2] Whipp and Phelps, p. 12.

[3] Whipp and Phelps, p. 12.

[4] Whipp and Phelps, p. 13.

[5] Bruce A. Macdonald to Hope Morritt, 15 June 1965.

Harry Prince

[1] Lauriston, *Lambton*, p. 174.

[2] Lauriston, *Lambton*, pp. 174–175.

[3] Lauriston, *Lambton*, p. 174.

4 "Lynch Law at Petrolia," *Sarnia Observer Advertiser*, 4 September 1866, p. 2.

5 Cheryl MacDonald, "Canada's Secret Police," *The Beaver* (June/July 1991), p. 44.

6 Cheryl MacDonald, p. 44.

7 Bruce A. Macdonald to Hope Morritt, 15 June 1965.

The Petrolia Assemblies
1 Whipp and Phelps, p. 28.

2 Helen Corey to Hope Morritt, 17 September 1992.

3 Bruce A. Macdonald to Hope Morritt, 15 June 1965.

4 Arnold Thompson to Charlie Whipp, 16 April 1984.

5 Whipp and Phelps, p. 42.

6 "Preserving Your Friends," *Sarnia Observer Advertiser*, 30 August 1867.

7 Grant, p. 11.

8 Phelps, p. 59.

9 "Petrolia Oil Report," *Sarnia Observer Advertiser*, 30 July 1862, p. 2.

10 Grant, p. 11.

11 Grant, p. 11.

Corporate Ventures
1 Phelps, p. 96.

2 Cronin, p. 25.

3 Phelps, p. 96.

4 Whipp and Phelps, p. 10.

5 Grant, p. 18.

6 Grant, p. 9.

7 "Jacob Lewis Englehart," *The Canadian Annual Review of Public Affairs*, 1916, Supplement (Toronto, 1917), p. 815.

8 Robert Page, "The Early History of the Canadian Oil Industry, 1860-1900," *Queen's Quarterly* (Winter 1984), pp. 854-855.

9 Whipp and Phelps, p. 14.

10 Grant, p. 3.

11 Page, p. 859.

12 Page, p. 859.

13 John G. Taylor to Hope Morritt, 23 July 1991.

Petrolia Family Politics
1 Phelps, p. 65.

2 Phelps, p. 112.

3 Charles O. Fairbank to Hope Morritt, 4 May 1991.

4 "Jacob Lewis Englehart," *The Canadian Annual Review of Public Affairs*, p. 816.

5 Whipp and Phelps, p. 14.

6 Whipp and Phelps, p. 66.

7 *Petrolia Advertiser*, 11 February 1881.

8 Helen Corey to Hope Morritt, 17 September 1992.

9 Whipp and Phelps, p. 25.

10 Grant, p. 32.

11 Charlie Whipp to Hope Morritt, 11 July 1992.

12 Newell, p. 35.

13 Whipp and Phelps, p. 13.

14 The Marchioness of Dufferin, p. 39.

15 Grant, p. 31.

Foreign Drillers and Imperial Dreams
1 Whipp and Phelps, p. 69.

2 Whipp and Phelps, p. 9.

3 Whipp and Phelps, p. 70.

[4] Whipp and Phelps, p. 67.

[5] Grant, p. 25.

[6] Phelps, p. 123.

[7] Phelps, p. 16.

[8] Phelps, p. 125.

[9] Grant, p. 29.

[10] Jacob Lewis Englehart, Biography, p. 2.

[11] Grant, p. 12.

[12] Englehart Biography, p. 2.

[13] Herring, "The Granting of Bonuses," *Petrolia Advertiser*, 18 June 1878.

[14] Silver Star Refining Company Advertisement, *Petrolia Advertiser*, 28 February 1879.

[15] *No Stone Unturned: The First 150 Years of the Geological Survey of Canada* (Ottawa: Minister of Supply and Services, 1992), p. 7.

[16] *No Stone Unturned*, p. 7.

Founding Imperial Oil Company, Ltd.
[1] Lauriston, pp. 178-179.

[2] Grant, p. 31.

[3] Charles Oliver Fairbank, "The Petrolia Story," *Northwestern Oil Journal* (1955), p. 7.

[4] Phelps, p. 62.

[5] Ewing, vol. 2, p. 69.

[6] Ewing, vol. 2, p. 62.

[7] *The Story of Imperial Oil* (Toronto: Imperial Oil Limited, 1991), pp. 4-5.

[8] Ewing, vol. 2, pp. 61-62.

[9] *The Story of Imperial Oil*, pp. 4-5.

[10] Ewing, pp. 64-67.

[11] Frederick T. Rosser, *London Township Pioneers* (Belleville, Ontario: Mika Publishing Company, 1975), pp. 67-68.

[12] Ewing, vol. 1, p. 63.

The Oil Exchange
[1] Ewing, vol. 2, p. 26.

[2] Whipp and Phelps, p. 25.

[3] Phelps, p. 96.

[4] Phelps, p. 100.

[5] Phelps, p. 151.

[6] "H. A. Fairbank's Funeral," *Petrolia Advertiser*, 11 February 1881.

[7] Phelps, p. 113, quoting letter from Edna Fairbank to Charles Fairbank, 22 November 1874.

[8] Phelps, p. 167.

[9] Phelps, p. 173.

[10] Phelps, pp. 173-74.

[11] Phelps, p. 174.

[12] "Almost a Tragedy," *Petrolia Advertiser*, 1 January 1892.

William McGarvey in Galicia
[1] Ed Gould, *Oil: The History of Canada's Oil and Gas Industry* (Victoria: Hancock House Publishers Ltd. 1976), p. 113.

[2] "Edward McGarvey and Son, Sale of Oil Lands," Advertisement, *Petrolia Advertiser*, February and March 1879.

[3] E. G., "Die Geschichte des Galizischen Erdoles," *Zeitschrift des Internationalen Verelnes der Bohringenieure*, 15 December 1914.

[4] Whipp and Phelps, p. 70.

[5] Martin Pollack, *Nach Galizien* (Wien-Munchen: Edition Christian Brandstatter, n.d.), p. 35.

[6] Dr. Ladislaus Szajnocha, The Petroleum Industry of Galicia (Vienna: Author's own publication, 1899), p. 14.

Imperial Success
[1] *The Story of Imperial Oil*, pp. 5-6.

[2] Ewing, vol. 2, pp. 70-71.

[3] Ewing, vol. 2, p. 68.

[4] Ewing, vol. 2, pp. 71-72.

[5] Ewing, vol. 2, p. 77.

[6] Sclanders, p. 3.

[7] *The Story of Imperial Oil*, pp. 6-7.

[8] Barry Broadfoot and Mark Nichols, *Memories: The Story of Imperial's First Century as Told by Its Employees and Annuitants*, Toronto (Don Mills, Ontario: Plow & Watters Printing, 1980), p. 2.

[9] Page, p. 856.

[10] Newell, p. 40.

[11] Page, p. 856.

Baronial Buildings
[1] Sclanders, p. 4.

[2] Joel Englehart, Death Notice, *The New York Times*, 12 August 1887.

[3] Petrolia Census, 1881.

[4] Whipp and Phelps, pp. 22-25.

[5] Phelps, p. 225.

[6] "A Millionaire's Showplace: Ghosts of Oil-Rich Petrolia Linger Within These Walls," *London Free Press*, 8 November 1959.

[7] Phelps, p. 225.

[8] Petrolia old timer to Hope Morritt, October 1992.

[9] "The Wedding," *Petrolia Advertiser*, 1 January 1892.

[10] "The Wedding."

[11] "The Wedding."

[12] "The Wedding."

Death of the Fathers
[1] Hamilton Cemetery Records, 26 November 1890.

[2] Phelps, p. 249.

[3] Phelps, p. 250.

[4] Phelps, p. 251.

[5] Phelps, p. 252.

[6] Phelps, p. 252.

Standard Oil Raids
[1] Page, p. 857.

[2] Ewing, vol. 3, pp. 72, 74.

[3] Ewing, vol. 3, p. 58.

[4] Whipp and Phelps, p. 59.

[5] Ewing, vol. 4, p. 78.

[6] Ewing, vol. 4, p. 79.

[7] Whipp and Phelps, p. 59.

Royal Petrolian
[1] Griffin, p. 23.

[2] Josef Muck, *Der Erdwachsbergbau in Boryslaw* (Berlin: Verlag von Julius Springer, 1903), p. 30.

[3] E. G., "Die Erdolgewinnung in Galizien," *Zeitschrift des Internationalen Verelnes der Bohringenieure und Bohrtechniker* — Nr. 8, 7 February 1914.

[4] C. Engler und H. Hofer, *Das Erdol, seine Physik, Chemie, Geologie, Technologie und sein Wirtschaftsbetrieb*(Leipzig: Verlag von S. Hirzel, 1909), pp. 324-325.

[5] *Petrolia Advertiser*, 19 December 1895.

[6] *Petrolia Advertiser*, 19 December 1895.

[7] *Petrolia Advertiser*, 19 December 1895.

[8] Henry James Morgan, ed., *Canadian Men and Women of the Time* (Toronto: William Briggs, 1912), p. 689.

Standard Takeover

[1] *The Story of Imperial Oil*, p. 7.

[2] Ewing, vol. 4, p. 85.

[3] Page, p. 864.

[4] Page, p. 864.

[5] John T. Saywell, "Through Famine to Fortune: The White Rose Story," Petrolia File, Lambton County Library Headquarters, Wyoming, Ontario, p. 7.

[6] Page, p. 860.

[7] Saywell, p. 7.

[8] Page, p. 863.

Ghost Town

[1] "A Millionaire's Showplace," *London Free Press*, 18 November 1959.

[2] *Petrolia Advertiser*, 12 March 1896.

[3] Phelps, p. 248.

[4] Michael Barnes, *Jake Engelhart* (Cobalt, Ontario: Highway Book Shop, 1974), pp. 1-2.

[5] "Jacob Lewis Englehart," Biographic Sketches, Lambton County Library Headquarters, Wyoming, Ontario.

[6] "Jacob Lewis Englehart," Biographic Sketches.

[7] "Jacob Lewis Englehart," Biographic Sketches.

Englehart's Railway

[1] Barnes, p. 30.

[2] Barnes, p. 32.

[3] Charlotte Eleanor Englehart, Last Will and Testament, 31 October 1908, The Petrolia Discovery Archives, Petrolia, Ontario.

[4] "Jacob Lewis Englehart," *The Canadian Annual Review of Public Affairs*, 1916 Supplement (Toronto, 1917), p. 818.

[5] "Jacob Lewis Englehart," Biographic Sketches.

[6] "Jacob Lewis Englehart," Biographic Sketches.

The End of an Era

[1] *Petrolia Advertiser*, 11 February 1914.

[2] Phelps, p. 273.

[3] "Late J. H. Fairbank Buried at Petrolia," *Sarnia Observer*, 13 February 1914.

[4] Franz Hubmann, *The Hapsburg Empire: The World of the Austro-Hungarian Monarchy in Original Photographs 1840-1916*, ed. Andrew Wheatcroft (London: Routledge and Kegan Paul, 1971), p. 343.

[5] E. G., "Zum 70. Geburtstage von W. H. McGarvey," *Zietschrift des Internationalen Verelnes der Bohringenieure und Bohrtechniker*, 7 July 1913.

[6] H. U., "W. H. MacGarwey," *Zeitschrift des Internationalen verlnes der Bohrtechniker und Bohringenieure*, Nr. 24, 15 December 1914.

[7] Barnes, p. 45.

[8] Barnes, p. 45.

[9] Sclanders, p. 7.

[10] Sclanders, p. 7.

[11] "Imperial Oil Pioneer Dies," *Imperial Oil Review* (1921), p. 11.

[12] "Jake Englehart: Petrolia's Amazing Benefactor," *Petrolia Advertiser-Topic*, 5 December 1963.

[13] Phelps, pp. 145, 182.

AFTERWORD

[1] *The Story of Imperial Oil*.

ILLUSTRATIONS

SOURCE /PAGE